589th ENGINEER BATTALION (CONSTRUCTION)
ENGINEERS AT WAR IN VIETNAM

An Operational Summary

1967 - 1971

To the Brothers We Lost

To Our Brothers Who Were Wounded – Emotionally and Physically

To Our Families Who Suffered

and

To Those who Returned

©2014 589th Engineer Battalion Association (Vietnam) An incorporated tax exempt organization under the laws of North Dakota.

Edition 1: September 2014

This book was co-authored by Dennie L. Pendergrass (Company B), Historian, 589th Engineer Battalion Association (Vietnam), and Larry W. Jinkins (Company C). Perry Blanchfield (Company A) provided immeasurable assistance by making pictures from the 589th Website available for this book. Bob Spencer (Company A) provided significant content suggestions.

Please direct any questions, content suggestions, or corrections to dennie@589thengineers.com.

Editing by Two Mules Editing, Von Pittman

Published by AKA-Publishing
Columbia MO

ISBN 978-1-936688-81-4

TABLE OF CONTENTS

- iii   Glossary of Acronyms
- v   Introduction
- vii   Historical Background
- viii   Map: 589th Area of Operations
- 1   Part 1: 589th Organization
- 5   Part 2: ORLL Synopses pertaining to personnel, attachments, missions and movements, and enemy actions.
- 63   Part 3: Operational, Organizational, and Logistical Issues and Lessons
- 73   Part 4: Awards and Citations
- 77   Part 5: Those We Lost
- 81   Part 6: Photographic History

GLOSSARY OF ACRONYMS

ARVN	Army of the Republic of Vietnam
BN	Battalion
Co	Company
CONUS	Continental United States
CPT	Captain
CMP	Corrugated Metal Pipe
CSM	Command Sergeant Major
CY	Cubic yards
DBST	Double Bituminous Surface Treatment
ENG	Engineer
HHC	Headquarters and Headquarters Company
LCU	Landing Craft, Utility
LF	Lineal feet
LOC	Lines of Communications (i.e. roads, airfields)
LST	Landing Ship, Tank
LT	Lieutenant
LTC	Lieutenant Colonel
MAJ	Major
MCA	Military Construction Army program
MTOE	Modified Table of Organization & Equipment
NVA	North Vietnamese Army
ORLL	Operational Reports—Lessons Learned
PLT	Platoon
ROKA	Republic of Korea Army
SFC	Sergeant First Class
SGT	Sergeant
SSG	Staff Sergeant
SF	Square feet
SY	Square yards
TOE	Table of Organization and Equipment
TPH	Tons per hour
VC	Viet Cong

589th ENGINEER BATTALION (CONSTRUCTION)
ENGINEERS AT WAR IN VIETNAM

18th Engineer Brigade

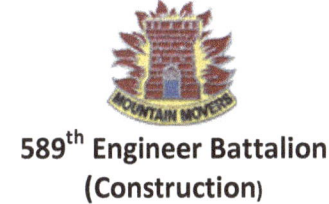

589th Engineer Battalion (Construction)

INTRODUCTION

The 589th Engineer Battalion (Construction) provided construction support for the U.S. Military in the Republic of Vietnam (RVN) from April 1967 until April 1971. Elements of the Battalion operated in the area from Qui Nhon to Pleiku from the Battalion's arrival in Vietnam until July 1968, as well as from the Phan Rang Air Force Base west towards Dalat, north towards Cam Ranh, and south towards Vinh Hao, from July 1968 until April 1971. A number of its construction projects are still in use by the Vietnamese today, including National Highway 19 between Qui Nhon and Pleiku, National Highway 27 (formerly QL-11) between Phan Rang and Dalat, and National Highway 1 from Phan Rang to Vinh Hao.

While much has been written about combat operations in Vietnam, little has been written about the "Engineer War." The most authoritative book to-date that describes the contributions engineers made during the conflict was published by the U.S. Center of Military History, and is entitled: *Engineers at War: The United States Army in Vietnam* (2011). While this book provides an amazing amount of detail about the engineer missions in Vietnam, it does not provide a "feel" for the daily life and the challenges of engineer units and engineer soldiers in Vietnam.

This synopsis of the 589th Engineer Battalion's quarterly "Operational Reports – Lessons Learned" (ORLLs) and information from other historic sources has been created to provide a feel for the daily life and challenges of an engineer unit in Vietnam. Much more detail is available in the full ORLLs. This synopsis may soon have a companion book, a compilation of short stories written by members of the 589th. The "Guestbook" section of the 589th Engineer Battalion Association

(Vietnam), website: http://www.589thEngineers.com, is also available to the public for a look into today's thoughts of our 589th veterans. It also includes the full ORLLs in its "History" section.

This synopsis is intended for use by:

- Former 589th members who desire a greater understanding of their involvement in the Vietnam War, or wish to clarify their memories;

- The families of 589th members and of other engineer unit members, who want to gain a better understanding of the involvement of their engineer fathers and grandfathers in the Vietnam War; and,

- Military historians who want an account of the everyday operations and life of a U.S. Army engineer unit at war in an environment of insurgency.

HISTORICAL BACKGROUND

Construction battalions typically operated behind the front lines in all previous wars, including World War II and the Korean War. The greatest enemy threats during those wars were from air attack and enemy artillery fire. On occasion, direct contact with enemy soldiers took place during breakthroughs, such as the one that occurred during the Battle of the Bulge. But direct contact with enemy soldiers was relatively rare for construction battalions. Also, during earlier wars, construction battalions typically operated across relatively short distances and in closely defined geographic areas.

Enemy contact for construction engineers in Vietnam differed significantly from that experienced in previous wars. As described in "Enemy actions" in Part 2, in Vietnam the enemy conducted extensive guerilla warfare in construction battalions' areas of operations. The North Vietnamese Army and the Viet Cong initiated mining incidents, ambushes, sniper fire, and mortar and rocket attacks on camps. Further, construction battalions in Vietnam often operated over much greater distances than had been the case in previous wars.

Meanwhile, the construction battalions of the Vietnam War went to war with Tables of Organization & Equipment (TOEs) that had been designed for conventional wars. *(Author's Note: A TOE is a document that states what a unit is authorized in the way of equipment and people, and how they were organized. An insurgency war conducted over long distances, based on a TOE designed for conventional wars, created unique challenges for engineer units in Vietnam, as evidenced in the following sections of this synopsis.)*

PART 1

589TH ORGANIZATION

The 589th arrived in Vietnam as organized under Engineer TOE 5-115E; by 1969 it was organized under TOE 5-115G; and by 1970 it was organized under the modified MTOE 5-115GPO5. The differences between these TOEs are not sufficient to warrant discussion in this document. The following account is a generalized description of the Battalion throughout its operations in Vietnam.

The Battalion was organized into five companies: Headquarters & Headquarters Company (HHC); one Engineer Equipment and Maintenance Company (Company A); and three "line" construction companies (Companies B, C, and D)

The Battalion had an authorized personnel strength that ranged from 935 to 1156, depending upon the TOE and authorization level at any given time. The TOEs authorized the Battalion to carry 30 to 31 commissioned officers and seven warrant officers.

HHC: This was called a Headquarters & Headquarters Company because it had two headquarters within it: the Battalion command-level headquarters with a Lieutenant Colonel as commander; and the Company command-level headquarters with a Captain as commander. The Company level headquarters was responsible for providing logistical support (housing, food, security, maintenance, supplies, mail distribution, etc.) to the Battalion headquarters.

The Battalion headquarters consisted of the Battalion Commander, a Sergeant Major, a Chaplain, and a number of sections including: S1 (Personnel), S2 (Intelligence), S3 (Operations), S4 (Logistics), Materials Testing, Surveying, Medical, and Communications. The Medical Section consisted of a doctor and medics. Typically one or more medics were attached to a company when it was not physically located near the Battalion aid station. Later during its time in Vietnam, the Battalion added a Civilian Personnel Office section that coordinated the hiring and pay of Vietnamese workers.

In addition to providing logistical support to the Battalion headquarters, the Company Headquarters also managed two water purification units that provided water for the Battalion and other military units, and a Utilities Section that constructed housing and defensive structures in support of the Battalion headquarters, and occasionally the other companies.

Company A: This Company was responsible for:

- Direct Support maintenance for all companies. Direct Support maintenance, also called Third Shop, entailed major maintenance work, such as replacing engines, transmissions, etc.;
- Contact maintenance, usually consisting of a truck and two mechanics who went to construction work sites to provide equipment repairs;
- Machine shop facilities;
- Rock crushing to produce aggregate for use in concrete and road and airfield base course;
- Bituminous dust palliation and limited asphalt paving; and,
- Specialized construction equipment operations in support of the construction companies.

The Company was organized into a Headquarters Platoon, a Construction Equipment Section, an Asphalt Section, a Rock Processing and Quarry Section, a Direct Support Maintenance Section, and a Support Section.

Companies B, C, D: Each of these companies had a Headquarters Platoon responsible for providing food, water, mail, administration, supplies, maintenance, company operations coordination and communications, and for coordinating religious and medical services. Each company had two "line vertical construction" platoons—First and Second Platoons—that did trades work such as carpentry, plumbing, electrical, and masonry. And, each company had a "horizontal construction" Earthmoving Platoon.

The mission of the First and Second Platoons was: "To construct and rehabilitate buildings, port facilities, bridges, drainage structures, pipelines, storage tanks and related facilities, and to install and repair utilities."

The mission of the Earthmoving Platoons was: "To excavate, haul, compact, and grade earth and to provide stabilized earth subgrades for airfields, roads, railroads, levees, supply storage areas, and similar projects."

Some of the major items of Battalion equipment included;

- Crawler Mounted 12 1/2-ton Crane Shovel – two each
- Truck Mounted 20-ton Cranes – six each
- Bituminous Distributor – two each
- Dump Trucks 5-ton – forty-eight each
- Ditching Machines – two each
- Graders – nine each
- Scoop Loaders 2 ½ cubic yard – nine each
- Concrete Mixers – six each
- Rock Drilling Equipment – one each
- Towed Scrapers 18 cubic yard – twelve each
- Dozers – fourteen each
- Water Purification Sets – two each
- Rock Crushing and Screen Plant – one each
- Pile Drivers – three each

PART 2

ORLL SYNOPSES

The following is a synopsis of the available 589th Engineer Battalion's quarterly Operational Reports - Lessons Learned (ORLLs) and information from other historic sources pertaining to personnel, attachments, missions and movements, and enemy actions. Much more detail is available in the full ORLLs, which are available to the public at http://www.589thEngineers.com – the 589th Engineer Battalion Association (Vietnam) website.

(Author's Note: The ORLLs only listed the assignments of officers. While the 589th Engineer Battalion Association (Vietnam) has obtained many of the company rosters that include the names of enlisted soldiers, many of them are substantially unreadable.)

1 MAY 67 - 31 JULY 67

Personnel

Staff and Line Officers at the time of movement to the Republic of Vietnam (RVN):

- Battalion Commander: Lieutenant Colonel Myron D. Snoke
- Battalion Executive Officer: Major Kenneth J. Davidson
- S-1: CPT Lawrence (Larry) Doff
- S-3: MAJ Edwin Marcy
- S-4: CPT Luis Alvarez
- Battalion Maintenance Officer: CPT Hugh Boyd III
- Battalion Signal Officer: 2LT Richard Schallenberg
- Battalion Chaplain: CPT Charles H. Walcott
- Battalion Surgeon: CPT William E. Jackson
- Sergeant Major: CSM Steven M. Dymezenski

- Headquarters & Headquarters Company Commander: Captain Charles E. Briggs (Commander)
- Company A Commander: CPT Edgar A. Marshall, 2LT Robert L. Sigle, 2LT Julian P. Robinson, CW3 C. W. Britt
- Co B Commander: CPT John A. Sparks, 1LT Billy R. Ellsbury, 1LT Gerald F. Zinngrabe, 2LT Dennie L. Pendergrass, 2LT Edward M. Fennell, CW2 William T. Kiely
- Co C Commander: CPT Elmer J. Lybert, 1LT James L. Campbell, 1LT Steve M. Albers, 2LT Thomas R. Englehart, 1LT David L. Sysco, CW3 Samuel Cooper
- Co D Commander: CPT David V. Harbach, 1LT Alfred P. Cochran, 2LT Edward H. Hegmann II, 2LT David W. Britain III, 2LT Tom Sanders, WO1 John A. Miller

Deployment

The 589th was activated on 21 Jan 66 at Fort Hood, Texas. After a year of training, inspections, and receipt of personnel and equipment to bring the unit to authorized levels, the Battalion received a Movement Directive on 12 Dec 66, with Movement Orders following on 15 Dec 66.

The Battalion moved equipment and supplies by train to Pier III City Docks, Beaumont, Texas. It loaded them, along with escort personnel, onto two ships, the Sheldon Lykes and the Ruth Lykes, for shipment on 25 and 26 Mar 67 respectively. The ships were unloaded in Qui Nhon on 1 May 67 and 5 May 67.

The Advance Party departed Fort Hood by plane on 17 April 67, and arrived at Qui Nhon on 19 Apr 67.

On 8 and 9 Apr 67, the main body of troops departed by rail for the Oakland Army Terminal to board the United States Naval Ship General John Pope (T-AP-110). The General John Pope departed on 11 Apr 67. After a brief stop at Subic Bay, Philippines, it arrived at Qui Nhon, Viet Nam on 29 April 67.

The Battalion deployed at nearly full personnel strength, but short several major pieces of authorized engineer equipment, including five of seven 250-CFM pneumatic tool and compressor outfits, four of six 20-ton truck mounted cranes, one of two ten-ton crawler cranes, and all six water distributors.

Attachments

- Upon arrival in Qui Nhon, the 589th was assigned to the 45th Engineer Group, 18th Engineer Brigade, APO San Francisco 96307.
- The 84th Engineer Battalion, located in Qui Nhon, was designated as host battalion to assist the Advance Party and the main body with staging-in.
- On 6 June 67, the 70th Engineer Company (Dump Truck) was attached to the 589th.

Detachments

- Immediately upon arrival in country, Co B, 589th was detached from the 589th and attached to the 84th Engineer Battalion. It began building a company cantonment site near Qui Nhon. Co B would later be permanently reassigned from the 589th to the 84th.
- On 24 June 67, one platoon from the 70th Engineer Company (Dump Truck) was detached from the 589th and attached to the 35th Engineer Battalion.

Mission and Movements

During the first week in-country, the Battalion was staged at Cha Rang Maintenance Depot, while a cantonment site was cleared six miles to the west at Cu Lam Nam. On 6 May the Battalion (minus D Co which moved to Vinh Thanh on 8 May and C Co which remained at the Cha Rang Maintenance Depot Cantonment) moved to its new Cu Lam Nam site. It began to build its cantonment area while taking over seven projects from other engineer units. They included route maintenance of 39 miles of QL-19, bridge rehabilitation, and construction of a 200,000 square yard Logistics Depot. In addition, the unit assembled pre-engineered metal buildings for a General Support Maintenance Facility and aircraft hangars.

The Battalion also received nine design and construct directives, including:

- orders for a CV2 Airfield at Vinh Thanh (construction began 8 May 67 and was completed 15 July 67);
- drainage structures and base course preparation of the An Khe pass,
- four bridges totaling 511 feet;
- fifty-two miles of base course preparation on routes QL-19, TL-3A and TL-6B;
- the 400-Bed Republic of Korea Army (ROKA) 6th EVAC Hospital;
- three well water fill sites; and,
- a Post Exchange (PX) Depot site to include 10 buildings, hardstands and roads.

Enemy Action

None reported.

1 AUGUST 67-31 OCTOBER 67

Personnel

On 1 Oct 67, MAJ Eddie L. Morris assumed duties as Executive Officer. This position had been vacant since 26 Aug 67, when former XO MAJ Kenneth J. Davidson was reassigned.

On 18 Oct 67, LTC Myron D. Snoke was reassigned to the 18th Engineer Brigade after fourteen-and-a-half months in command of the 589th. On 24 Oct 67, LTC Allen F. Grum assumed command of the Battalion.

On 23 Oct 67, the following major staff and command positions became effective:

- 1LT Alfred P. Cochran, formerly construction officer of Co D, assumed command of HHC from CPT Lawrence D. Doff.
- CPT George Thein, formerly Battalion Civil Engineer, assumed command of Co A from CPT Edgar A. Marshall.
- CPT Lawrence D. Doff, formerly HHC Commander, assumed command of Co D from CPT David V. Harbach, who assumed duties of the Battalion Civil Engineer.

- CPT Edgar A. Marshall was reassigned from Co A to S-4.
- CPT John S. Hillmer was reassigned from S-4.
- CPT Thomas T. Takayama, formerly Battalion Pipeline Engineer, assumed command of the 511th Eng. Co (Panel Bridge) from 2LT Henry V. Soper on 30 Oct.

Attachments

- Company B, 84th Engineer Battalion was attached to the 589th on 10 Oct 67. It remained at Camp Radcliff, An Khe. On 23 Oct 67, this Company would be permanently assigned to the 589th as Co B. In exchange, the 589th's Co B was permanently reassigned to the 84th.
- The 51st Engineer Platoon (Asphalt) was attached to the 589th on 22 Sept 67.
- The 23rd Engineer Detachment (Well Drilling) was attached to the 589th on 28 Oct 67.
- The 444th Engineer Detachment (Concrete Mixing and Paving) was attached to the 589th on 10 Oct 67.
- The 511th Engineer Company (Panel Bridge) was attached to the 589th on 10 Oct 67.

Missions and Movements

On 10 Oct 67, the Battalion's area of responsibility was extended west by approximately twenty-five miles. It reached from the top of the An Khe Pass to the base of the Mang-Giang Pass. Its responsibilities included upgrading and maintaining QL-19, performing bridge repairs, and construction projects within the confines of Camp Radcliff, An Khe. As a result of this increased commitment, Co D, a section of Co A's third echelon maintenance shop and quarry personnel, relocated to Camp Radcliff, as did half of the battalion's operations section.

HHC: The Utility Section upgraded the Battalion base camp by improving drainage, erecting forty-five utility poles to accommodate installation of 100 floodlights, improving shower facilities, and installing hot water heaters. The section also constructed over 500 international road signs for the various battalion road projects. The Water Point Section, using two erdalator units, provided water to an average of fifteen company and battalion size units by treating and issuing 2.4 million gallons of potable water. One of HHC's unique projects was the design and fabrication of a road paint-striping machine that was eventually used to stripe 40 miles of roads. This machine drew considerable attention from higher level Engineer Staffs in all sectors of the country. It became a regular stop for "Command Inspections."

Co A: On 7 Aug 67, A Company's Quarry Section operation, consisting of one 75TPH (ton per hour) primary crusher and one 75TPH secondary crusher, moved from Vinh Than to the An Son Valley. After setting up in An Son, the unit began operating an additional 75TPH primary unit and screening plant from the 73rd Engineer Company. From 7 Aug to 31 Oct, A Company produced approximately 115,000 cubic yards of crushed rock to meet requirements for a Double Bituminous Surface Treatment (DBST) at the Logistics Depot, the Post Exchange Depot, and the Korean ROKA Hospital. It also supplied rock for headwalls and floor slabs for concrete projects east of the An Khe Pass. On 10 Oct 67, the unit received two more 75TPH primary crushers, plus a cone crusher and two secondary crushers, at An Khe. After performing required maintenance, the An Khe site produced nearly 11,000 cubic yards of crushed rock—primarily base course rock—to meet aggregate needs for Battalion projects in the An Khe area. The Asphalt Section remained heavily committed on continuous requirements for DBST work on Battalion Projects.

Co B: On 10 Oct 67, this unit was detached from the 70th Engineer Battalion and attached to the 589th. It remained at An Khe. During this period the Company continued work on three major projects, including a 4,365-foot long concrete airstrip. This was a twenty-four-hour-per-day operation, with base rock preparation during daytime and concrete placement at night. Co B completed the airstrip on 10 Sept 67; it became operational on 27 Sept 67. Traffic used taxi turn-outs while permanent facilities were under construction. A main Post Exchange facility with a 120-foot x 200-foot concrete pad and a Pascoe type building neared completion, with only interior air-condition duct work installation remaining.

Another unit completed the main skeletal steel work of a maintenance hangar project, and Co B installed all roofing, siding, and additional bracing. The remaining work required placing 750 cubic yards of reinforced concrete flooring and installation of twelve large sliding doors and electrical wiring.

Co C: The Company remained at the Cha Rang Maintenance Depot. It completed the Logistic Depot project, consisting of 200,000 square yards of DBST treatment, 40,000 square yards of surfaced roads, and numerous drainage structures. Work continued on six General Support Maintenance facility buildings. The Post Exchange depot complex neared completion, with four of eight closed storage 40-foot x 200-foot Pascoe buildings and two 40-foot x 200-foot open storage buildings completed. In addition, 47,000 square yards of hardstand received a DBST treatment, and 970 lineal feet of culvert corrugated metal pipe were installed. On road LOC TL-6B 7, Co C upgraded six kilometers of road, from intersection of QL-2 and 6-B to the Republic of Korea Army Tiger Division entrance. This required the placement of 33,000 cubic yards of base course rock. To date, two kilometers of this road received 90,000 square yards of DBST. The company upgraded three bridges on QL-19, 1,000 lineal feet of 24WF (wide flange) steel beams with heavy timber decking. Twelve-pile timber bents were constructed on two of the bridges.

Co D: Construction of the 400-bed ROKA (Republic of Korea Army) 6th Evacuation Hospital in the An Son Valley continued. This hospital will consist of seven air-conditioned Quonset huts, built on concrete pads with two-foot concrete block sidewalls, covering an area of 12,480 square feet. These buildings will accommodate admissions, an X-ray facility, pre- and post-operative wards, two operating rooms, an intensive care ward, and a morgue. The hospital will also have six single-story buildings constructed of tropical-wood sides, built on concrete pads and covering 10,000 square feet. These six buildings will house a pharmacy, a records and administrative area, a supply facility, an EENT (eyes, ears, nose and throat) clinic, an officers' lounge, and a nurses' quarters complex. The electrical distribution system, water distribution, and sewage system have been installed. Upgrading QL-19 in the An Khe Pass involved the repair and cleanout of seventy culverts and the installation of nine additional multi-barrel culverts with headwalls. Co B repaired two bridges, using 1160 feet of 36WF230 steel beams and heavy timber decking. One bridge required construction of a twelve-pile timber bent. Over 8,000 cubic yards of crushed-rock base course material were placed in preparation for a hot mix asphalt surface. Upgrading the LST (Landing Ship, Tank) beach in Qui Nhon included placing 30,000 cubic yards of laterite soil, river run rock, and crushed rock with a single bituminous surface treatment.

51st Engineer Platoon (Asphalt): The Platoon arrived in country minus the following equipment: one 150 TPH asphalt plant, two paving machines, two 8 ton rollers, one 10 ton roller, and one 2 ½ ton water distributor. The expected arrival dates for this equipment are unknown. The unit was employed mainly on DBST work. It used its dump trucks to support the battalion's haul requirements. Unit personnel are being used to operate the additional two primary crushers.

70th Engineer Company (Dump Truck): During this period, the Company supported the 589th, the 35th Engineer Battalion (Combat), and for a short period the 19th Engineer Battalion (Combat). In its support missions, the unit committed an average of 45 trucks daily, seven days per week, including both day and night operations. Unit personnel and drove over 425,000 miles. Company-supported projects included rehabilitation of QL-19 and QL-1, bridge construction on QL-1, a logistics depot project, and rock crusher support. In addition, the unit completed construction of its cantonment area.

23rd Well Drilling Detachment: This Detachment arrived with two enlisted men, but minus all TOE equipment, during the last week of this period. The men have been further attached to Co A of the 589th.

444th Engineer Detachment (Concrete Mixing & Placing): From 1 Aug to 10 Sept 67, the Detachment worked around the clock mixing and paving for the 4,365 linear foot An Khe Army Airfield. They mixed and placed more than 10,000 cubic yards of concrete. From 11 Sep to 31 Oct the unit produced concrete for base development projects, including warehouses, ammunition pads, headwalls, and billet pads.

511th Engineer Company (Panel Bridge): The Company used its resources to perform its secondary mission as a dump truck company, deploying an average of twenty-five trucks daily. Its major hauls consisted of moving material to the concrete mixing plant in support of the concrete airstrip construction. The unit performed minor reinforcements for existing Bailey bridges on QL-19. It supported the construction of culverts, timber trestle bridges, helicopter pad revetments, and the access road on Camp Radcliff. The unit also operated a prefabrication yard employing eighty local national Vietnamese, under the supervision of three enlisted soldiers, for the construction of buildings for units on Camp Radcliff.

Enemy Actions

None reported

1 NOVEMBER 67 – 31 JANUARY 68

Personnel

No staff or officer changes were listed in this ORLL.

During this period the Battalion underwent a rotational hump alleviation program. Over a two-week period a total of 130 persons were exchanged with other units within the 45th Engineer Group.

Attachments

- 511th Engineer Company (Panel Bridge)
- 51st Engineer Platoon (Asphalt)
- 444th Engineer Detachment (Concrete Mixing and Paving)

Missions and Movements

This quarter covered the monsoon season, characterized by a marked slowdown in construction efforts due to rain. Fortunately, this area was not affected to the extent that construction efforts ever came to a complete standstill for any length of time.

HHC: The Utility Section continued to upgrade the Battalion base camp by installing an additional nine culverts with headwalls, erecting six guard towers, and constructing an eight room, 20-foot x 64-foot dispensary. Daily Vietnamese hires fortified the S-3 Operations tent with sandbags and 55-gallon drums filled with sand. A Battalion theater was framed and covered with canvas, enabling its use during the rainy season. HHC installed a concrete pad for the company Motor Pool maintenance shop and hot water heaters in the mess hall. The Water Purification Section produced nearly 250,000 gallons of potable water, serving an average of fifteen company and battalion size units.

Co A: The Quarry Section moved half of its men and part of its equipment to An Khe from the Republic of Korea (ROK) Valley in October. By 1 Nov 67, the Hon Cong quarry site at An Khe was in full production, as was the ROK Valley site. Between the two sites, over 36,000 cubic yards of base course rock were produced during the month of November, as well as nearly 3,400 cubic yards of concrete aggregate, DBST rock, and fines. By 1 Dec 67, the entire Quarry Section had moved to An Khe and begun full-scale production. Over 16,000 cubic yards of base course was crushed during December while operating only one crusher for twelve hours per day. During January more than 27,000 cubic yards of base course were produced, along with 2,600 cubic yards of DBST and fines. A multi-bench quarry was excavated to provide material to the three primary crushers.

The Asphalt Section used material from these quarries to work on QL-1, QL-19 and Route LTL-6B. The Battalion general support maintenance facility and parking lot were given a DBST surface. The Section hand-sprayed over 16,000 gallons of RC-3 at the An Khe petroleum tank farm to protect the tank berms. Working with Co C, more six miles of road were given a sand-asphalt treatment in less than a week.

The Equipment Section was fully committed supporting other sections, as well as operations in other companies. Company personnel used lowboy trailers to haul equipment and supplies for S-4, and to haul the dozers and front loaders that supported the quarry and asphalt sections, as well as other battalion projects.

Co B: During this quarter the Company remained at Camp Radcliff, An Khe. It engaged in four major construction projects within the confines of the cantonment. They included work on the airfield's connecting and parallel taxiways and access road that complemented the recently completed concrete runway, a 190-foot x 175-foot maintenance hangar, two fifty-four foot airfield control towers, and the Camp Radcliff power distribution system. The Main Post Exchange and the berm stabilization at the 65,000-barrel tank farm were completed.

Co C: As the only construction unit of the battalion in the area of responsibility east of the An Khe Pass, this unit's responsibilities and workload increased considerably during the quarter. Projects included the Cha Rang Post Exchange Depot, thirty-one helicopter revetments, and a stabilized refueling pad. Further, the Company completed base preparation of the Qui-Nhon interim access road at DeLong Pier and the construction of 160 feet of permanent bridging, four stabilized firing platforms for a 175mm howitzer battery, and the erection of panel bridging.

Co D: The Company was fully committed to projects in the An Khe area of operations, including completion of a 400-bed hospital for the Republic Of Korea Army (ROKA) in the An Son Valley. It upgraded Route QL-19 from the An Khe pass to the base of the Mang Giang Pass and Route TL-3A from the junction of QL-19 to Vinh Thanh. Co D constructed the 1st Calvary Division Supply Point yards at Camp Radcliff, two taxiways and revetments on the "Golf Course" heliport. It conducted frequent mine sweeps on QL-19 and TL-3A.

51st Engineer Platoon (Asphalt): During the first two months, the Platoon engaged in DBST work in support of Co C. In late December the asphalt plant arrived. By 7 Jan 68, site preparation was under way at Hon Cong Mountain, in An Khe, for the asphalt plant.

511th Engineer Company (Panel Bridge): During this period the unit acted primarily under its secondary mission as a dump truck company and logged over 133,250 miles in support of battalion projects. Under its primary mission, the unit made extensive repairs to, replaced wear treads on, and upgraded several panel bridges along QL-19.

444th Engineer Detachment (Concrete Mixing and Paving): Monsoon rains hampered site preparation for concrete placement. The unit's production dropped from slightly over 10,000 cubic yards for the two preceding quarters to less than 3,000 cubic yards for this quarter.

Organization Day: On 21 January 1968, the unit celebrated the second anniversary of its activation at Fort Hood with an "Organization Day." Suitable programs were conducted at both Cu Lam Nam and An Khe. They included company competitions, picnic fare, and professional entertainment.

Civic Action: During this period, the Battalion was recognized by the Commanding General of the Qui Nhon Support Command for having made outstanding contributions to the Civic Action Program during the month of December.

Enemy Actions

- Thirteen mining incidents were reported in the 589th area of operations. Seven occurred on Route TL-3A during the time that Co D was upgrading it. Two Co D vehicles were involved in the minings, resulting in the wounding of three soldiers.

- Two attacks occurred on QL-19 bridges. The first attack occurred on 6 Dec 67 on bridge QL-19-12. The second attack occurred on 27 Jan 68 on bridge QL-19-11. Both bridges sustained damage and were subsequently repaired by the 589th.

- During an enemy breach of the Camp Radcliff perimeter, the 511th suffered one casualty in action during an enemy breech of the perimeter.

1 FEBRUARY 68 – 30 APRIL 68

This period included "Tet 68"

Personnel

During this reporting period, four of the five companies experienced turnover of their commanders:

- On 21 March 1968, CPT Alfred Cochran, HHC Commander, departed for CONUS.

- From 21 March to 4 April, 1LT Thomas Englehart commanded HHC. He in turn was replaced by 1LT James Montague, who also continued to serve as Communications Officer.

- On 3 April 1968, CPT Phillip Suitt replaced CPT Marcel Reynolds, as Commander of Co B. CPT Reynolds moved to the position of S-4, replacing CPT Edgar Marshall, who returned to CONUS.

- On 4 April 1968, 1LT John W. Morgan assumed command of Co C, replacing CPT James Campbell, who was reassigned to Ft Belvoir.

- On 12 April 1968, CPT Theodore Adkins replaced CPT Lawrence Doff in command of Co D. CPT Doff returned to CONUS.

During this reporting period the Battalion had an average enlisted strength 132 men below its authorized level, due to rotations to CONUS. During the first ten days of April, approximately 280 enlisted men rotated. During the entire quarter, while 18 officers departed, only 6 arrived.

Attachments

On 15 March 1968, the 589th Engineer Battalion was reassigned from the 45th Engineer Group to the 35th Engineer Group. Units attached to the 589th at this time included:

- 51st Engineer Platoon (Asphalt)

- 511th Engineer Company (Panel Bridge)

- 444th Engineer Detachment (Concrete Mixing & Paving)

Detachments

- On 15 Feb 68, the 511th Panel Bridge Company was detached from the 589th and attached to the 45th Engineer Group.

- On 20 Mar 68, the Co C Earthmoving Platoon, supported by mechanics from A and C Companies and an HHC Medic, was placed under the operational control of the 45th Engineer Group. It was subsequently attached to the 14th Engineer Battalion (Combat) located in Quang Tri Province.

- On 9 Apr 68, Co B was placed under the operational control of the 937th Engineer Group (Combat).

Missions and Movements

HHC: The Utilities Section continued the tasks of enhancing living conditions and improving support capabilities at the base camp. It constructed a 10-foot x 24-foot shower, an additional 20-foot x 52-foot tent billet, and a grease rack. It placed a 20-foot x 32-foot concrete slab in the maintenance tent. Several projects improved perimeter defense, including emplacing 6,200 feet of double- apron barbed wire fence, 12,400 feet of concertina fence, and thirty-nine trip flares. HHC erected an additional guard tower, dug an 81mm mortar pit, and performed continuous maintenance on sixty-three perimeter defense structures. The water point operation was transferred to another unit on 23 Feb 68, due to the possibility of unit deployment, after 650,000 gallons of potable water had been produced during the period.

Co A: The Company continued to provide maintenance support to QL-19 through the use of asphalt equipment and aggregate production. The asphalt plant began operations on 8 Apr 68. In order to produce aggregate for the plant, Co A established two types of crusher complexes. The initial crusher setup for this aggregate production consisted of a 75 TPH primary unit (jaw crusher), a cone crusher, and a secondary unit (roll crusher). Initial runs with this operation resulted in a failure to meet gradation specifications, due to the large corrugations on the roll crusher. An additional secondary unit was modified by installing two smooth rolls with welded beads parallel to the long axis. Adding this machine to the complex provided four stages of rock size reduction. It also enabled the crushing of coarse aggregate to specifications. It was learned early in this operation that sufficient fine aggregate could not be produced to operate the asphalt plant on a daily basis. The complex was converted to fine aggregate production and an additional crusher complex was set up to produce coarse aggregate for the asphalt plant.

Quarry production for the period totaled 30,700 cubic yards of 2-inch minus base course material, 3,200 cubic yards of 1-inch minus coarse aggregate, 3,800 cubic yards of fine aggregate, and 2,160 cubic yards of two-inch minus concrete aggregate.

The Direct Support Engineer Maintenance shop overhauled four rock crusher units and 252 items of engineer equipment. It repaired 136 items of ordinance equipment. These efforts accounted for a total of 8,742 man-hours.

Co B: During the first two months of the reporting period, the Company was engaged in four major projects within the confines of Camp Radcliff, An Khe. They included work on the An Khe Airfield, the construction of two control towers (fifty-two feet each), the construction of aircraft maintenance hangars, and the installation of a power distribution system.

On 9 April 1968 the Company moved to Pleiku and was placed under the operational control of the 937th Engineer Group (Combat). At Pleiku the Company initiated the construction of two 80-foot x 144-foot aircraft maintenance hangars, a 20,000 KW (kilowatt) power generating station, and a forty-foot x 100-foot pre-engineered steel warehouse. It placed 35,600 square yards of M8A1 matting for hardstands and upgraded QL-14 north of Pleiku, toward Dak To.

Co C: The Company continued to work on projects in the Cha Rang area, An Son Valley (ROK Valley), and Highway QL-19 from the junction of Highways QL-19 and QL-1 to the foot of An Khe Pass. The Company was designated as an infantry reaction force for the Qui Nhon area during "Tet 68." It secured the airport from 7 Feb through 4 Mar 1968. On 20 Mar, the Earth Moving Platoon was placed under the operational control of the 45th Engineer Group north of Hue.

Co D: The Company continued to upgrade QL-19 from An Khe to the base of Miang Giang Pass. It began reconstruction of bridge 19-26, and started construction of an access road up Hon Cong Mountain to a signal facility. Projects completed during the period included the Division Supply Point, access roads, and the Division heliport. Additionally, the Company assumed responsibility for Company B projects at An Khe.

51st Engineer Platoon (Asphalt) began paving operations on QL-19 west of An Khe on 8 April. Two items delayed the start of paving. First, not all items required to set up the asphalt plant had arrived. Secondly, due to problems with the rock crusher, sufficient quantities of aggregate that would meet gradation specifications were not available.

511th Engineer Company (Panel Bridge) continued to train replacement personnel and to perform their secondary mission of dump truck support. On 15 February 68, the Company was detached from the 589th and attached to the 45th Engineer Group (Construction).

444th Engineer Detachment (Concrete Mixing and Paving): Concrete production was decreased during the reporting period due to lessened demand and the lack of sufficient earthmoving equipment to prepare job sites.

Civic Action: The Battalion continued its contribution to the Civic Action Program by providing local, regional and popular forces (RF/PF) with construction material to build defense bunkers and reinforce perimeter areas. It distributed 5,000 pounds of surplus foodstuffs to dependents of local RF/PF, completed a five-room school in An Khe on 31 March 1968 (funded by the citizens of Waynesville, Ohio), provided men and equipment to complete the Binh Khe High School, and made monetary donations to the Phu Phong Orphanage. The Battalion Surgeon examined and treated an average of 1200 Vietnamese per month in the battalion dispensary and at two Vietnamese dispensaries.

Enemy Actions

- Eight mine incidents occurred on QL-19.
- Two road graders were lost in combat. On 21 Feb 68, one 5-ton dump truck received seven hits from enemy automatic weapons, resulting in the driver being hit with glass.
- On the night of 2 Feb 68, the motor pool and barracks for the 51st and A Co sustained a mortar attack, resulting in one serious injury requiring evacuation.
- On 3 Feb 68 a former Co D member, who had been reassigned to Co C, 84th Engineer Battalion, was killed on guard duty by wounds received when hit by fragments from a hostile explosive charge.
- On the night of 4 Mar 68, the billet area of Co A and D at Camp Radcliff, An Khe, sustained a mortar attack in which 14 personnel were injured.

1 MAY 68 – 31 JULY 68

Personnel

- On 26 May 68, CPT Michael E. Gilbertson (then 1LT) replaced MAJ Kenneth W. Ashley (then CPT) as A Company Commander.
- On 2 Jun 68, MAJ Kenneth W. Ashley (then CPT) replaced the Battalion S-3 Officer, MAJ Clyde E. Richmond.
- On 6 Jul 68, Battalion Commander LTC Allen F. Grum was replaced by LTC Albert C. Costanzo.
- On 11 Jul 68, 1LT Bruce E. Fernandez assumed command of HHC from 1LT James G. Montague .
- On 18 Jul 68, MAJ Leslie R. Wieduwilt replaced LTC Eddie L. Morris (then Major) as Battalion Executive Officer.
- On 22 Jul 68, 1LT John W. Morgan turned over the command of Company C to 1LT Herbert L. Hines.
- On 22 Jul 68, 1LT Jack A. Purcell assumed the vacant position of Engineer Equipment Maintenance Officer. On 3 Feb 68 a former Co D member, who

had been reassigned to Co C, 84th Engineer Battalion, was killed on guard duty by wounds received when hit by fragments from a hostile explosive charge.

- On 24 Jul 68, 1LT Lawrence A. Willwerth replaced CPT Marcel R. Reynolds as Battalion S-4.

Attachments

- 23rd Engineer Detachment (Well Drilling)
- 51st Engineer Platoon (Asphalt)
- 444th Engineer Detachment (Concrete Mixing and Placing)
- 542nd Engineer Detachment (Power Line)
- 614th Engineer Detachment (Power Line)
- 643rd Engineer Company (Pipeline) First Platoon

Detachments

- Co C Earthmoving Platoon continued to be attached to the 14th Engineer Battalion (Combat) north of Hue.
- The 23rd Engineer Detachment (Well Drilling) was detached from the 589th and attached to the 864th Engineer Battalion on 6 Jun 68.

Mission and Movements

HHC: The Utilities Section continued maintenance and upkeep of the base cantonment and assisted Co C with utility installation work. One water purification unit was pressed into service on 28 Jun 68 at the Battalion base camp, due to non-availability of a sufficient potable water supply from outside sources.

Co A: Maximum effort was expended on the upgrade and maintenance of Highway QL-19. The Asphalt Section performed road maintenance along 75 kilometers of pavement where heavy equipment had caused rutting of the surface and base damage, and broken pipelines had fed fires that had charred the pavement. Track vehicles had damaged bridge approaches while using bridge bypasses. The Quarry Section, comprised of three 75 TPH crusher units and a cone crusher, produced approximately 60,000 cubic yards of various size aggregate.

Co B: The Company was stationed at Pleiku. It participated in six major construction projects, including the construction of an aircraft maintenance hangar and a power generator plant, the placement of 32,000 square yards of M8A1 at the Kontum Aviation Facility, the upgrading of ten kilometers of QL-14 between Kontum and Dak To, and construction of ammunition supply point (ASP) berms at Kontum.

During the last month of the reporting period, Co B moved all organic personnel and TOE equipment from Pleiku to Phan Rang, beginning 10 Jul 1968.

Co C: During the reporting period, Company C, minus its Earthmoving Platoon, was located at the Cha Rang Maintenance Depot. It remained under the operational control of the 14th Engineer Battalion (Combat) and was stationed at various locations in the Quang Tri area. Earthmoving Platoon projects included constructing an aircraft parking apron at the Camp Evans airfield—completed on 14 Jun 1968, upgrading a thirteen-kilometer access road from Wunder Beach to QL-1 at Hai Lang, and extending the Mai Loc Airfield from 1800 to 2300 feet. *(Author's note: The 14th Combat Engineer Battalion History reports that volunteer members of the Battalion and attached units hauled ammunition into the Khe Sanh area during Operation Pegasus, in relief of the Marines surrounded at Khe Sanh, during the first week of April.)*

The two construction platoons completed Bridges 6B-34, 6B-33, and 6B-32, all of which required upgrading and approach widening. Three aircraft hangars at Lane Army Airfield were completed. Water and sewer lines, power distribution and a water tower with storage were constructed at Port Kamp Facility at Cha Rang Depot.

On 24 July 1968 Co C, minus its Earthmoving Platoon, began moving all its personnel and equipment to Phan Rang. It arrived on 26 July 1968. The Company staged in the Battalion area. On 28 Jul 1968, it began moving to Song Pha (forty-eight kilometers west of Phan Rang). The Earthmoving Platoon departed Wunder Beach via LST for Phan Rang on 16 Aug 68.

Co D: The Company was primarily involved in construction work on QL-19, construction on Camp Radcliff, and support of two tactical units in the area, civil affairs groups, and an ARVN Engineer Company. Primary bridgework involved the completion of bridges 19-26 and 19-24, and the repair of bridge 19-21. Upgrading and preparation of QL-19 for asphalting of a 19-kilometer section from LZ Scheuller to Mang Giang Pass was completed.

The Company supported tactical units along QL-19 by preparing a five-acre area for the construction of a new artillery firebase, LZ Action, near Bridge 19-27. It also prepared tank and bunker locations at all bridges from 19-18 to 19-27.

23rd Engineer Detachment (Well Drilling) was attached to the Battalion on 28 Oct 1967, upon arrival in-country without their well-drilling equipment. On 6 Jun 1968 they were detached, then attached to the 864th Engineer Battalion. Their equipment still had not arrived in country.

51st Engineer Platoon (Asphalt) was totally committed to paving QL-19 during this period. The unit produced 20,680 tons of asphalt and paved 18 kilometers of two-lane road. The two-inch pavement was applied in one lift.

444th Engineer Detachment (Concrete Mixing & Paving) was used to complete a Pasco, 80-foot x 216-foot pre-engineered metal aircraft hangar at Camp Radcliff during this period. The project was turned over to Co D in July 1968.

542nd and 614th Engineer Detachments (Powerline): These detachments were activated 5 January 1968, at Fort Belvoir, Virginia. After a period of training, the units boarded ships at Oakland, CA on 1 May 1968, and arrived at Cam Ranh Bay on 22 May 1968, where they were attached to the 589th Engineer Battalion. The units were sent to An Khe to finish installation of a power distribution system that required setting 172 poles, erecting 197 guys and anchors, emplacing 84 transformers, making 417 service drops from secondary voltage system facilities, and stringing 101,060 linear feet of wire.

First Platoon, 643rd Pipeline Company: This Platoon was placed under the operational control of the 589th on 25 July 1968, for the purpose of relocating certain sections of the fuel pipeline that ran from Qui Nhon to the air base at Phu Cat. The pipeline was located along a railroad and crossed twenty-seven railroad bridges. The project included offsetting the pipeline at all bridges, such that sabotage directed against the often-hit pipeline would not necessarily result in damage to the vital railroad bridges.

Newspaper: A Battalion newspaper, *The Mountain Mover News*, was established as a bi-weekly publication, with the first issue scheduled for publication on 1 August 1968.

Civic Action: All children from the Binh Khe Orphanage and their teachers and administration were periodically invited to the compound to share religious services and to eat lunch with the troops. They were given tours of the battalion base camp. The Battalion Surgeon administered to approximately 1500 Vietnamese under the Medical Civic Action Program (MEDCAP) at daily sick call.

Enemy Actions

- On 5 May 68, one Co D grader was destroyed by mine detonation. Three Co D personnel were wounded.

- On 5 May 68, bridge QL-19-21 was overrun by thirty to forty Viet Cong.

- On 25 May 68, bridges at QL-19-26 and QL-19-21 received mortar fire.

- Four vehicles (two 290M scrapers and two graders) belonging to the Earthmoving Platoon of Co C hit mines. One grader was destroyed; three grader operators were wounded. These incidents occurred in Quang Tri Province on the LZ Jane to Wunder Beach road. *(Author's Note: The mines were reported in a 14th Combat Engineer Battalion ORLL and in the book, Engineers at War, as being new Soviet mines that the U.S. Army's mine detectors could not detect.)*

- On 19 May 68, a section of the Co C Earthmoving Platoon was pinned down by flying shrapnel for fourteen hours at Camp Evans, when the ammunition depot and several fuel bladders exploded during an enemy attack. This incident was reported in the ORLLs of many units.

- A grader belonging to Co D struck a mine at Coordinates BR289454 on Route QL-19 on 5 May 1968. One man was seriously wounded and returned to CONUS. Two other men were slightly wounded and returned to duty after two weeks.

1 AUGUST 68 – 31 OCTOBER 68

Personnel

Five units of the battalion changed command during this period.

- On 7 Aug 68, CPT Richard Comiso assumed command of Co C from 1LT Herbert L. Hines.
- On 15 Aug 68, 1LT Robert P. Grant, 51st Engineer Platoon (Asphalt) Commander, departed.
- On 5 Sep 68, 1LT Arthur Davis assumed the vacant position of Communications Officer. On 12 Sep 68, 1LT Robert R. Greer assumed command of Co B from CPT Philip W. Suitt.
- On 17 Sept 68, Chaplain (CPT) James E. Rogers took over the Battalion Chaplain duties from Chaplain (CPT) Wendell O. Hawley.
- On 21 Sep 68, 1LT Steven Schilson assumed command of the 51st Engineer Platoon (Asphalt).
- On 29 Sep 68, CPT Joseph Feast Jr., former Battalion Civil Engineer, assumed command of Co D from CPT Theodore A. Adkins, who then became assistant S-3 of the 35th Engineer Group (Construction).
- On 29 Sept 68, CPT Thomas O'Dea became the Battalion Civil Engineer, replacing CPT Joseph Feast Jr.
- On 12 Oct 68, 1LT Charles S. Poteet replaced 1LT Jack Purcell as Engineer Equipment Maintenance Officer. 1LT Purcell became Platoon Leader for Earthmoving Platoon of Co C.

- On 13 Oct 68, 1LT David M. Swope assumed command of the 513th Engineer Company (Dump Truck) from 1LT Patrick W. Meyers.
- On 13 Oct 68, MAJ Richard B. Pierce assumed Battalion XO duties from MAJ Leslie R. Wieduwilt.
- On 23 Oct 68, CPT Lannie R. Hughes replaced CPT James A. Greco as Battalion Surgeon.
- On 26 Oct, CW4 Allen Keeney assumed duties of Supply Technician and Property Book Officer from CW4 Donald A. Drach.

Attachments / Detachments

- Co D - detached to the 84th Engineer Battalion from 13 Aug 68 to 15 Sep 68
- 513th Engineer Company (Dump Truck) - detached 17 Aug 68 and reattached 13 Oct 68
- 51st Engineer Platoon (Asphalt) - detached 1 Aug 68 and reattached 8 Oct 68
- 444th Engineer Detachment (Concrete Mixing & Placing) - detached 1 Aug 68
- 614th Engineer Detachment (Power Line) detached 1 Aug 68
- Earthmoving Platoon, Co C – reattached from the 14th Engineer Battalion to the 589th on 16 Aug 68.

Missions and Movements

HHC: On 27 Jul 68 the Company was alerted for a road march from Cu Lam Nam to Qui Nhon, and for an ocean move by LST from Qui Nhon to Phan Rang Air Force Base. On 3 Aug the Company moved to the Qui Nhon LST Beach, loaded and departed the beach on 4 Aug, and arrived at Phan Rang on 5 August, 1968.

Upon arrival, the Utilities Section immediately began improving and rebuilding the new Battalion base camp facilities located within the Phan Rang Air Force Base. Work accomplished included: rewiring and rescreening of two mess halls, twenty-five billets, one Battalion Headquarters and four Staff Section buildings;

cleaning and rebuilding three latrines; constructing a Battalion Conference Room and a basketball court; installing hot water heaters in three showers; improving mess facilities; wiring the 116th Engineer Battalion's cantonment area at Phan Rang Air Force Base; constructing unit and section signs; and constructing three bunkers for unit defense.

During this period, one water purification unit was attached to Company C at Song Pha. The other water purification unit initially processed water at Phan Rang, and then was moved to the Company D base camp at Don Duong on 27 Sept 68. Approximately 1,500,000 gallons of water were processed during this period.

Co A: The Company began moving to Phan Rang on 9 Aug 68. This move required several LSTs (Landing Ship, Tank), LCUs (Landing Craft Utility) and a seagoing barge. All Personnel and equipment arrived in Phan Rang by 22 Aug 68. A nine-man section of the Equipment Platoon was sent to Bac Loc on 21 Aug to repair a M8A1 runway. On 8 Oct a rock crusher was set up west of Phan Rang to begin crushing base course rock for future use on QL-11.

Co B: During this period, the Company was involved in five large construction projects, including upgrading and building bridges on a 14.4 mile stretch of QL-11 between Phan Rang and the Tan My Bridge, constructing a base camp for two companies of the 116th Engineer Battalion (Idaho National Guard) at Di Linh, placing ammunition revetments at the Phan Rang Air Base, constructing MACV facilities in the Phu Quy sub-sector, and performing maintenance on a seventy-five mile stretch of QL-1 from BN 253412 To BN 893947.

Co C: The Company constructed a new base camp cantonment area at Song Pha, maintained and upgraded QL-11 from BN 619957 to BP 453084, and supported the movement and establishment of a base camp for the 116th Engineer Battalion (Construction). The Earthmoving Platoon departed Wunder Beach on 16 Aug, via LST for Phan Rang, then moved to Song Pha via road march.

Co D: Primary activities involved completing construction projects at An Khe, moving the unit from An Khe to Don Doung (departed An Khe on 16 Sep 68, arrived at Phan Rang on 19 Sep 68, and began deployment to Don Duong on 23 Sep 68), constructing a new base camp at Don Duong, and maintaining QL-21A and QL-11.

51st Engineer Platoon (Asphalt): The unit supported Co A quarry operations and the Battalion's airfield maintenance and upgrading missions. The Platoon's paving train remained attached to the 19th Combat Engineer Battalion for paving at LZ English. Upon completion of that task, the paving train rejoined the Platoon.

513Th Engineer Company (Dump Truck): The unit supported paving operations on QL-19 through 17 Aug 68.

Enemy Actions

- On 2 Aug 68, a half-mile west of bridge QL-19-25, a 290M scraper struck a mine. The explosion destroyed the tractor and wounded the operator.
- On 3 Aug 68, a 5000-gallon tank trailer traveling east on QL-19 struck a mine, destroying the trailer's rear axle.
- On 20 Aug 68, enemy sappers damaged Q-11 bridges 6, 10, and 27.
- On 21 Aug 68, Phan Rang Air Base received an 82mm mortar attack with no casualties.
- On 5 Sep 68, an enemy demolition team destroyed QL-11 Bridge 13.
- On 11 and 15 Sep 68, Phan Rang Air Base received mortar fire with no casualties.
- On 14 Sep 68, the 1st Platoon of Co B was attacked at Di Linh. No 589th casualties; one enemy KIA.
- On 17 Sep 68, a 589th convoy traveling west on QL-21A to Bao Loc sustained automatic weapons and mortar fire. No 589th casualties.
- On 19 Sep 68, the Co C/ARVN Regional Force compound at Song Pha received 10 rounds of mortar fire. No 589th casualties.
- On 1 Oct 68, the culvert at QL-11 former bridge site 13 was destroyed, along with the immediately adjacent railroad bridge.

- On 8 Oct 68, a Co B vehicle was ambushed en route to TL-408 by three enemy soldiers. Resulted in light damage to the vehicle and no US casualties.

- On 30 Oct 68, the Co C/ARVN Regional Force compound at Song Pha received one rocket and two mortar rounds, resulting in four Vietnamese wounded and no US casualties.

1 NOVEMBER 68 - 31 JANUARY 69

Personnel

The Battalion and four units of the Battalion changed command during this period.

- On 4 Nov 68, 1LT Kenneth P. Koppers was assigned as Battalion Adjutant.

- On 13 Nov 68, 1LT Roger D. Langford assumed command of HHC from 1LT Bruce R. Fernandez.

- On 16 Nov 68, LTC Al S. Rosin assumed command of the Battalion from LTC Albert C. Costanzo, who was on emergency leave.

- On 20 Nov 68, 1LT Howard M. Baga was assigned as Civilian Personnel Officer (CPO).

- On 22 Nov 68, CPT Forrest P. Hanson was assigned as Assistant S-3.

- On 1 Dec 68, 1LT Bruce R. Fernandez assumed the duties of S-4 Officer.

- On 12 Dec 68, CPT Thomas O'Dea assumed duties of S-3 upon MAJ Ashley's departure.

- On 21 Dec 68, CPT Thomas O'Dea assumed command of Co D.

- On 21 Dec 68, CPT Joseph Feast Jr. assumed duties as S-3.

- On 7 Jan 69, 1LT Anthony C. Muse, Jr. assumed command of the 51st Engineer Platoon (Asphalt) from 1LT Steven Schilson, who became Construction Officer of Co B.

- On 28 Jan 69, CPT Forrest P. Hanson assumed command of the 513th Engineer Company (Dump Truck) from 1LT David M. Swope.

Attachments

The following organic and attached units comprised the 589th Engineer Battalion during this period: HHC, Co A, Co B, Co C, Co D, 513th Engineer Company (Dump Truck) and the 51st Engineer Platoon (Asphalt).

Missions and Movements

HHC was engaged in construction to significantly improve living conditions and defensive positions in the Battalion area. Water purification units and personnel were supplied to Co C and Co D, with a total of 410,000 gallons of water produced.

Co A: The Company's primary function during this period was to support the line companies with specialized equipment and maintenance. On 21 Nov 68, the Quarry Section received a 225 TPH crusher to replace a 75 TPH crusher to support paving operations in Don Duong, where Co D was located. During the period 37,400 cubic yards of base course and concrete aggregate were produced. The Maintenance Section received an average of ten work orders per day. Approximately sixty percent were for engineer equipment. The remainder were ordnance items.

Co B: Working out of Phan Rang, the Company was engaged in three large projects, which included building roads and upgrading bridges on a 14.4 mile stretch of QL-11 between Phan Rang and the Tan My Bridge, coordinating BN 812788 to BN 618957, upgrading MACV facilities in the Phu Quy Sub-Sector coordinate BN 742745, and performing road maintenance on a 75-mile stretch of QL-1 from coordinate BN 253412 to BN 893947.

Co C: The Company upgraded QL-11 from coordinates BN619957 to BN 599992, provided continuous maintenance for a 32 kilometer section of Route QL-11 from coordinates BN 619957 to BN 453084, completed its base camp at Song Pha, and provided support to the 5/27th Artillery by constructing a base camp area—known as Fire Base Flo-Anne—for two 155 mm howitzers and their gun crews. Fire Base Flo-Anne was located on the west side of Bridge 16. The Company also completed the extension of bridge 38.1 (BP 45860927), which eliminated one of the most dangerous curves on QL-11.

Co D: The bulk of the work centered on road maintenance and on reconditioning the road surface on QL-11 from Dalat to a point 7.5 miles east of Don Duong and all of QL-21A from Don Duong to QL-20. The Company also repaired a blown French Eiffel bridge at BP 299028 on QL-21A. Two new projects were started in Dalat at Can Ly Airfield. One involved repairing a weak area at the end of the runway. The other consisted of building four POL tanks, a distribution line and access roads in the POL area.

51st Engineer Platoon (Asphalt): The major activities were supporting Co A with tractor-trailer and dump truck haul capability and preparing the asphalt plant for use in the near future.

513th Engineer Company (Dump Truck): This Company was attached to the 589th Engineer Battalion but under Operational Control of other engineer battalions.

Enemy Actions

- On 13 Nov 68, the railroad bridge near bridge QL-11-50 was destroyed.
- On 25 Nov 68, bridge QL-21A-6 was destroyed.
- On 26 Nov 68, a 1/4 ton truck belonging to Co B—while traveling north on QL-1—was destroyed, with two 589th personnel killed and one wounded. *(Author's Note: the sole survivor of this ambush, Al Carlisle, writes about this incident and his subsequent avoidance of the enemy while wounded in his book,* Depth of Field; An Army Photographer's Year in Vietnam.*)*
- On 29 Nov 68, a convoy of Co D and Vietnamese Regional Forces troops exchanged small arms fire with enemy forces at bridge QL-21A-6.
- On 3 Dec 68, the Co C supply truck was ambushed near bridge QL-11-9. The Company Supply Sergeant was killed, and a company cook received five small arms fire wounds, but survived following extensive hospitalization and evacuation to CONUS.

- On 16 Dec 68, a Co D work party observed abnormal behavior by civilians near the hamlet of Bac Binh. MACV and Regional Forces were notified. Upon entering the village they made contact with a Viet Cong platoon.

- On 18 Dec 68, a Co C vehicle travelling about six miles east of Song Pha on QL-11 struck a command-detonated mine. The vehicle received minor damage and the driver received minor injuries.

- On 23 Dec 68, elements of Co C received small arms fire near bridge QL-11-19. No US casualties.

- On 30 Dec 68, an element of Co B received five rounds of small arms fire one kilometer west of bridge QL-11-5. No casualties.

- On 4 Jan 69, a Co C dozer operating near bridge QL-11-17 uncovered an anti-personnel mine.

- On 7 Jan 69, an element of Co C found a mine while working west of bridge QL-11-19.

- On 14 Jan 69, the Co C base camp at Song Pha received mortar fire resulting in the death of one Earthmoving Platoon soldier. Two were wounded. One storage building and five vehicles were damaged.

- On 15 Jan 69, Co D guards on the base camp perimeter received five rounds of small arms fire. No casualties.

- On 16 Jan 69, the culvert bypass at bridge QL-11-22 was destroyed.

- On 26 Jan 69, Phan Rang Air Base received mortar fire. There was slight damage to A Co motor pool equipment.

1 FEBRUARY 69 – 30 April 1969

Personnel

- CPT Grant L. Fredericks assumed command of Co B, replacing 1LT Robert R. Greer, who rotated.

- On 1 Feb 69, CW2 Harold D. Mangum became the personnel officer, replacing CW2 James Kessinger, who rotated.

- On 12 Feb 69, CPT Joseph P. Leary became Adjutant, replacing 1LT Kenneth P. Koppers.

- On 12 Feb 69, 1LT James D. Creaseman, Commander 553 Engineer Company (Float Bridge) was assigned to the Battalion.

- On 13 Feb 69, 2LT Charles (Chuck) Schueddig was assigned as Platoon Leader of 2nd Platoon, Co C, replacing 1LT Timothy Collins, who was assigned as the Battalion S3 Construction Engineer.

- On 26 Feb 69, 1LT Kenneth P. Koppers became Commander, HHC, replacing 1LT Roger Langford, who rotated.

- On 26 Feb 69, MAJ Martin M. Warvi assumed the S-3 duties, replacing CPT Joseph Feast Jr., who became Assistant S-3.

- On 8 Mar 69, CPT Richard W. Chapman assumed command of Co A, replacing CPT Michael E. Gilbertson, who was assigned to the S-3 section prior to rotating.

- On 10 Mar 69, MAJ Eugene W. Bennett became Battalion XO, replacing MAJ Richard B. Pierce, who rotated.

- On 15 Mar 69, 1LT Anthony C. Muse and the 51st Engineer Platoon (Asphalt) were reassigned to the 577th Engineer BN.

- On 1 Apr 69, CPT Howard H. Reed and his 687th Engineer Company (Land Clearing) were attached to the 589th.

- On 1 Apr 69, CSM Clifford Moore succeeded CSM Frank Leeder, who rotated.

- On 22 Apr 69, 1LT Gerald N. Bradley assumed duties of Engineer Equipment Officer replacing 1LT Charles Poteet, who was reassigned to B Co.

Reassignments

- On 13 Mar 69, CPT Thomas O'Dea and Co D of the 589th were reassigned to the 577th Engineer Battalion.
- On 13 Mar 69, 1LT Robert A. Hixon and Co D of the 577th were reassigned to the 589th Engineer Battalion.

Attachments

- 51st Engineer Platoon (Asphalt)
- 513th Engineer Company (Dump Truck)
- 533rd Engineer Company (Float Bridge)
- 687th Engineer Company (Land Clearing)

Missions and Movements

HHC continued to improve living conditions in the battalion base camp area through light construction and minor repairs. The water point supporting Co C at Song Pha produced 368,000 gallons during the period and the one supporting Don Duong produced 351,000 gallons through 15 March, when it returned to Phan Rang.

Co A continued to perform its primary mission of supporting the line companies with specialized equipment and maintenance services. The paving operation began laying asphalt on 23 Feb. On 24 Feb, the quarry section began a partial switch to asphalt aggregate production. During the reporting period, 61,600 cubic yards of base course aggregate, concrete rock, and asphalt aggregate were produced. The Company Maintenance Section received an average of seven work requests per day.

Co B: On 1 Ap the Earthmoving Platoon of Co D, commanded by 1LT Samuel Lamey, was attached. Major projects during this period included: road upgrading and drainage structures along a 15.5 mile stretch of QL-11 between Thap Cham and the Tan My bridge, coordinates BR 767823 to BN 619957; road clearing and maintenance on a 102 mile stretch of QL-1; from coordinates CP 023304 to BN

182408; upgrading MACV facilities in Phu Quy, Buu Son and Tuy Phong subsectors; fire base facilities construction for the 6/84 Artillery at Tan My Bridge; and billet revetment construction for that portion of the 589th located at Phan Rang Air Base.

Co C: During this period the Company upgraded QL-11 from coordinates BN619957 to BP579033 and maintained Route QL-11 from coordinates BN619957 to BP 453084 (32 kilometers of roadway). The Company improved its cantonment area, completed a base camp area for two 155mm howitzers and their gun crews, in support of the 6/84th Artillery; and, upgraded MACV advisor facilities at Du Long.

Co D: (1 Feb – 12 Mar 1969). The bulk of this period's work consisted of road maintenance of QL-11 from Dalat to a point 7.5 miles east of Don Duong, and all of QL-21A from Don Duong to highway QL-20. Operational support was provided at Pr'line Mountain and Lang Bian Mountain signal sites, along with continued work at Cam Ly Airfield and C Battery 5/27 Artillery in Dalat.

(13 Mar – 30 Apr 1969) During this period, the newly assigned Co D from the 577th (less the EM platoon which remained at Tuy Hoa attached to Co B, 577th Engineer Battalion) convoyed from Tuy Hoa to Dong Ba Thin. On 31 Mar 69 the Platoon boarded an LST at the Vung Ro Bay harbor and was shipped to Phan Rang, where it was attached to Co B, 589th to provide support in upgrading QL-11. When the Company arrived at Dong Ba Thin, it started work on a counter-mortar site, an aircraft direct fueling facility at the Dong Ba Thin airfield and a warehouse for the 608th Transportation Company.

51st Engineer Platoon (Asphalt): Major activities for this period included maintenance of equipment, preparation for its move to Don Duong, and setting up the asphalt plant at that location.

513th Engineer Company (Dump Truck): The Company was attached to the Battalion for administration for all but eleven days of this period. The Battalion received one platoon for work on 19 April 1969.

533rd Engineer Company (Float Bridge): The unit was active with bridging missions during the early weeks of the reporting period, and with preparation for relocating the Company from Phu Hiop to Dong Ba Thin. The Company was relocated primarily to provide emergency bridging of the gap between the mainland and the Cam Ranh Bay peninsula, should the enemy destroy the My Ca Bridge located there. Construction sites, abutments and approaches were prepared and material stockpiles were established to perform the mission, should it become necessary. The unit was most active in performing its secondary mission of transportation after the move to Dong Ba Thin.

687th Engineer Company (Land Clearing): The primary mission of the Company during this period was to clear 200 meters along QL-20 from the II-III Corps border to QL-21, all of QL-21A and QL-11 from Song Pha to Phan Rang, and to perform general clearing around the Qui Nhon and An Khe areas. On 14 Feb 69, the entire Company was located at Camp Smith and attached to the 116th Engineer Battalion (Idaho National Guard).

Enemy Actions

- On 8 Feb 69, two culverts were destroyed on the QL-11 Good View Pass mountain road between Song Pha and Don Duong.

- On 22 Feb 69, members of a ARVN battalion killed seven Viet Cong and captured one near culvert 19 on QL-11.

- On 22 Feb 69, the Viet Cong burned forty percent of the village Luong Tri, just outside of the Phan Rang Air Base.

- On 24 Feb 69, the Phan Rang Air Base came under mortar and rocket attack with no friendly casualties.

- On 1 Mar 69, a Co B 10-ton truck was damaged by a Claymore mine in the Good View Pass. One soldier was wounded.

- On 4 Mar 69, the bypass culvert at bridge QL-11-20 was damaged by enemy demolitions.

- On 8 Mar 69, the bypass culvert at bridge QL-11-20 was again damaged by enemy demolition. The culvert was damaged twice more during the week of 9-13 Mar.

- On 15 Mar 69, two HHC vehicles drew fire near bridge QL-11-6. No casualties.

- On 15-24 Mar 69, Phan Rang Air Base came under mortar and rocket attacks on 5 occasions. No 589th casualties.

- On 25 Mar 69 at 1910, the Co C base camp at Song Pha came under mortar attack. Three soldiers from 2nd Platoon were wounded.

- On 17 Apr 69, Co C received sniper fire at culverts 19 and 20 on QL-11. No casualties.

- On 21 Apr 69, Phan Rang Air Base came under attack with no casualties.

- On 21 Apr 69, the large culvert (number 20 on QL-11), consisting of five large barrels was demolished by enemy explosives.

1 MAY 69 – 31 JULY 1969

Personnel

- On 17 May 69, 1LT Perry L. Price assumed command of HHC from 1LT Kenneth P. Koppers, who was reassigned to Co B.

- On 4 Jun 69, LTC Donald A. Ramsay became Battalion Commander, replacing LTC Al S. Rosin.

- On 4 Jun 69, 1LT Anthony C. Muse became the S-4 Officer, replacing 1LT Bruce Fernandez.

- On 15 Jun 69, CPT Joseph Feast Jr. became the S-3 Officer, replacing MAJ Martin M. Warvi, who rotated.

- On 25 Jun 69, 1LT Jack A. Purcell from Co C departed for CONUS.

- On 7 Jul 69, MAJ Eugene W. Bennett, the Battalion XO, was reassigned to the 35th Engineer Group as S-3. The XO position remained vacant for the remainder of this period.

- On 9 Jul 69, CPT James A. Lewis assumed command of Co D, replacing CPT John R. Logue, who was reassigned as Battalion Adjutant.
- On 9 Jul 69, 2LT David G. Bennett was assigned as Battalion Pipeline Engineer.
- On 9 Jul 69, 1LT Samuel L. Lamey was transferred from Co D to assume command of Co C from CPT Richard Comiso, who rotated on 25 July.
- On 25 Jul 69, CPT Frederick M. Howard assumed command of the 687th Engineer Company (Land Clearing) to replace CPT Howard H. Reed, who rotated.
- On 26 Jul 69, 2LT Donald E. Alsted became the Battalion Signal Officer, replacing 1LT Arthur M. Davis.

Attachments

- 51st Engineer Platoon (Asphalt)
- 513th Engineer Company (Dump Truck)
- 553rd Engineer Company (Float Bridge)
- 687th Engineer Company (Land Clearing)
- Earthmoving Platoon of Co C, 93rd Engineer Battalion, was attached on 6 July.

Missions and Movements

HHC: The Company operated a water point at Co C, Song Pha, and produced 638,000 gallons of water. The water point at Song Mao Airfield, supporting an element of Co, B produced 268,000 gallons. The Utilities Section continued to improve living conditions in the battalion area through light construction and minor repairs. The Section also began construction of fifty writing desks, using scrap lumber as it became available, for the Buddhist school near Phan Rang, as a Civic Action project.

Co A: The 51st Engineer Platoon (Asphalt) remained attached to the Company to assist in the battalion paving operation. Civilian contract mechanics were assigned to the Company on 3 Jun 69 for maintenance and repair of MCA-LOC commercial construction equipment procured to augment the TOE equipment. During this period the Company paved 24.4 kilometers of single-lane asphalt pavement on QL-11 between Phan Rang and Song Pha. Nearly 103,000 cubic yards of aggregate were produced even though production capability was hampered by crusher maintenance problems. *(Author's Note: MCA is the acronym for Military Construction Army funds. In this context, it was a program to procure commercial construction equipment to be maintained by contract civilian mechanics. LOC is the military term for Lines of Communication. To engineers, this meant roads and airfields.)*

Co B: On 15 June the Co D Earthmoving Platoon was detached from Co B and returned to Co D. Major projects during the period included: road upgrading along a 12.6 mile stretch of QL-11 between Thap Cham and Tan My Bridge, clearing and maintenance of a 102 mile stretch of QL-1, construction of MACV facilities at Phu Quy subsector, and constructing a base camp for the 2/1 Calvary Squadron at Song Mao and Vinh Hao. Earth moving related projects were also provided for several Civic Action projects.

Co C: The Company upgraded and maintained thirty-two kilometers of QL-11 and making ongoing improvements to its cantonment area. A continual maintenance program was established to insure that QL-11 was kept open to traffic. The Second Platoon assisted with construction of the Co D base camp at Phu Quy from 29 Jun to 17 Jul. Work included constructing seven live-in/fight-in bunkers and assistance with mess hall construction.

Co D: The Company was involved in construction of an operations bunker in Cam Ranh Bay, transferring or completing projects in the Dong Ba Thin area, moving the unit from Dong Ba Thin to Phan Rang, constructing the Company base camp at Phu Quy, and upgrading of QL-1 from Phan Rang to Xom Moi . On 11 Jun, the Company (less the 2nd Platoon) began the move to Phan Rang Air Base and on 22 Jun it moved to the new camp at Phu Quy. On 6 Jul the EM Platoon from Co C of the 93rd Engineer Battalion was attached to assist in upgrading QL-1.

513th Engineer Company (Dump Truck): During this period the Company hauled rock and asphalt for construction and paving operations on QL-11.

553rd Engineer Company (Float Bridge): While attached to the 589th, the unit was most active in performing its secondary mission of transportation, which included general hauling and supporting relocation of the 547th Asphalt Platoon, Co B, 577th Engineer Battalion and Co D, 589th.

687th Engineer Company (Land Clearing): The primary mission during this period was to clear along QL-11 from Song Pha to Phan Rang, to conduct general clearing operations in the Phan Thiet area, and to assist with clearing during Operation Pipestone Canyon.

Enemy Actions

- On 16 May 69, the Co A paving crew received small arms fire near the village of Trai Nho Ho. No casualties.
- On 27 May 69, elements of Co B received small arms fire while camped near the village of Tuy Phong. No casualties.
- On 6 Jun 69, the Co C base camp at Song Pha came under mortar attack. No casualties.
- On 7 June, a former member of Co A, who had been reassigned to Co E, 1st Battalion, 52nd Infantry, 198th Light Infantry Brigade, was killed by a booby trap.
- On 19 Jun 69, elements of Co B received 5-6 rounds of 60mm mortar fire at Vinh Hao.
- On 26 Jun 69, 687th dozers uncovered a Viet Cong bunker/tunnel complex near the village of Binh Lam.
- On 27 Jun 69, 687th dozers detonated two booby-trapped 105mm rounds near the village of Binh Lam.
- On 30 Jun 69, 687th dozers discovered an eight-bunker complex near Binh Lam.

- On 1 Jul 69, 687th dozers uncovered a five-bunker complex near Binh Lam.
- On 4 Jul 69, the 687th uncovered twelve VC bunkers near LZ Sherry.
- On 18 Jul 69, the Co D base camp at Phu Quy received four 82mm mortar rounds.
- On 21 Jul 69, the Co D base camp at Phu Quy received one 60mm mortar round.
- On 28 Jul 69, elements of Co B received small arms fire while working on QL-11 near Dong Me.
- On 30 Jul 69, elements of Co D spotted a VC squad near the village of Thon Hieu Thien.

1 AUGUST 69 – 31 OCTOBER 69

Personnel

- On 1 Aug 69, CPT Alan J. Arikian assumed duties of Engineer Equipment Maintenance Officer (EEMO), replacing CPT Gerald N. Bradley, who rotated.
- On 4 Aug 69, MAJ Vincent Parmesano filled the vacant Battalion XO position.
- On 4 Aug 69, 1LT David R. Johnson was assigned as Second Platoon Leader in the 513th. 1LT Richard A. Peace continued as First Platoon Leader.
- On 5 Aug 69, CW3 Frank C. Cheatum was assigned as Maintenance Officer of Co D.
- On 14 Aug 69, CPT Alan J. Arikian vacated the Engineer Equipment Maintenance Officer position to assume command of the 513th Engineer Company (Dump Truck), replacing CPT Forest P. Hanson who returned to CONUS.
- On 16 Aug 69, CPT Daniel Quintard assumed command of Co B, replacing CPT Grant B. Fredericks, who was reassigned as an Assistant S-3 at the 35th Engineer Group.
- On 1 Sept 69, CW4 Allen Keeney, Unit Supply Technician, rotated to CONUS.

- On 7 Sep 69, CPT Robert Kyle assumed command of Co A, replacing CPT Richard W. Chapman who rotated to CONUS.

- On 12 Sep 69, WO1 Percy C. Green arrived to fill the Unit Supply Technician position.

- On 12 Sep 69, CPT James Rogers departed.

- On 13 Sep 69, MAJ Romayne E. Schroder assumed duties of S-3 and CPT Joseph Feast Jr., became Asst S-3.

- On 16 Sep 69, CPT Michael Burkowsky was assigned as Pipeline Engineer, replacing 1LT David G. Bennett, who was reassigned to Co B as a Platoon Leader.

- On 16 Sep 69, WO1 Thomas D. Macon filled the Unit Personnel Technician position vacated by CW2 Harold Mangu, who rotated to CONUS on 24 July 1969.

- On 1 Oct 69, MAJ Arnold T. Ellsworth was assigned as Battalion Chaplain.

- On 1 Oct 69, 1LT Charles W. Schueddig assumed command of Co C for five weeks, while CPT Sam Lamey returned to CONUS on leave.

- On 13 Oct 69, 1LT Byron Smith was designated as the Pipeline Engineer to replace CPT Burkowsky, who assumed command of HHC, replacing 1LT Perry L. Price, who returned to CONUS.

- On 15 Oct 69, Doctor (CPT) Richard E. Lavigno became the Battalion Surgeon, replacing CPT Lannie R. Hughes, who rotated.

- On 17 Oct 69, 2LT Russell Bement III was assigned duties as EM Platoon Leader in Co D.

- On 20 Oct 69, CPT Ferdinand Scafuro was assigned to the S-3 section to replace CPT Feast, Assistant S-3.

- On 26 Oct 69, 1LT Albert W. Hunt, Co D Construction Officer and XO, departed. 1LT Terence C. Holland, Co D Earthmoving Platoon Leader moved into the company Construction Officer and XO positions.

- On 26 Oct 69, 1LT Timothy W. Collins, Construction Engineer, S-3 section, rotated.

- Co B lost to rotation 1LT Steven B. Schilson, 1LT Ralph T. Arnold, and 1LT Kenneth P. Koppers. Officer gains were 1LT David G. Bennett, 1LT James G. Goetz and WO1 Robert A. Brown.

Attachments

- 513th Engineer Company (Dump Truck)
- 687th Engineer Company (Land Clearing)

Missions and Movements

HHC: The Company continued to operate two water points: at Song Pha, which produced 556,000 gallons of water in support of Co C; and at Song Mao, which produced 426,000 gallons of water in support of the 2/1 Cav. It continued to perform its usual mission of supporting the Battalion Headquarters throughout the period.

Co A: The Company provided the battalion with equipment, quarry, maintenance and paving support throughout the period. It paved 10.3 kilometers of roadway, using 10,773 tons of asphalt obtained from the 554th Red Horse Squadron (Air Force) asphalt plant. The battalion provided the aggregate and liquid asphalt. Equipment problems severely curtailed aggregate crushing. The Direct Support Maintenance Platoon received 357 job orders during the report period, of which fifty-five are still open, twenty-five are awaiting parts, twenty are awaiting shop work, and ten are in the shop at the time of writing this report. A significant portion of the direct support effort was expended on crusher maintenance and repair. DYNALECTRON contract mechanics provided maintenance assistance for MCA-LOC equipment throughout the period.

Co B: Projects during this period included support missions for 2/1 Calvary Squadron in Song Mao; revetments for army helicopters at the Phan Rang Air Base; major road repair of QL-1 in Binh Thuan Province; and upgrading a 1.2- mile stretch of haul road. On 24 Sep 69 a provisional platoon of personnel from the construction and earthmoving platoons were deployed on QL-1 to repair damages caused by the monsoon rains. All remaining excavation and sub-base work on "DEROS HILL" (BN 692872) was completed.

Co C: The Company engaged in upgrading QL-11 from coordinates BP 540058 to BP 507075 and maintaining constant traffic along QL-11, from coordinates BN

619957 to BN 453084 (34 kilometers) through maintenance operations. Due to heavy monsoon rains during September and October, more concentrated efforts were applied to the Good View Pass between Song Pha and Don Duong to keep QL-11 open continuously.

Co D: Major Company effort involved upgrading QL-1 between Phan Rang and Ruong Muo (BN 726713 to BN 707606); clearing 229.9 acres from BN 719688 to BN 706660 for security purposes and maintaining a 33 kilometer section of QL-1 from BN 800771 to BN 654533. In addition, the Company performed emergency flood control work at the fire base for Q Troop, 2/1 Calvary Squadron. It continued work on the Company base camp and Civic Action projects in the Phu Quy District.

513th Engineer Company (Dump Truck): The Company continued to operate its two dump truck platoons in separate locations until 17 Oct, when the Second Platoon (formerly attached to Co A, 864th Engineer Battalion) rejoined the First Platoon at Phan Rang to assist in the upgrade of QL-11.

687th Engineer Company (Land Clearing): The First and Second Platoons engaged in clearing operations in Binh Dinh Province from 11 Aug through 7 Oct They also engaged in clearing operations along QL-1, out to 200 meters on both sides of the roadway between Phan Thiet and Luong Son.

Enemy Actions

- The Co D base camp at Phu Quy received mortar fire on six occasions (5 Aug, 12 Aug, 5 Sep, 14 Sep, 6 Oct, and 9 Oct), with no casualties.

- The Phan Rang Air Base came under mortar and rocket attacks on 5 Sep and 20 Sep. No casualties.

- On 8 Aug 69, the Co D Earthmoving Platoon working on QL-1 came under enemy rocket-propelled grenade and small arms fire for 20 minutes. No casualties. *(Author's Note: the ORLL appears to be incorrect in that it has been confirmed through official records that two Co D soldiers were killed on this day by mortar fire.)*

- On 14 Aug 69, the 687th was hit by small arms fire and mortar fire while in convoy vicinity of BN 025360. Casualties were light.

- On 21 Aug 69, a Co B work party at Song Mao received mortar and rocket fire. No casualties.

- On 2 Sep 69, the railroad bridge and the corresponding road culvert at bridge QL-11-9 were destroyed by the enemy.

- On 17 Aug – 7 Sep 69, the 687th found and destroyed 186 bunkers and 10 spider-type foxholes in the Lee Hong Phong Forest area.

- On 9 Sep 69, a Co D work party received small arms fire from three directions simultaneously while working on QL-1. No casualties.

1 NOVEMBER 69 – 31 JANUARY 70

Personnel

- On 9 Nov 69, CPT David Berkman assumed command of the 687th Engineer Company (Land Clearing) from CPT Frederick Howard.

- On 9 Nov 69, the Quarry Platoon of the 73rd Engineer Company (CS) arrived with Platoon Leader 1LT Gary Woodard.

- On 25 Nov 69, CPT Frank A. Robertson assumed command of Co D from CPT James M. Lewis, who rotated.

- On 4 Dec 69, 1LT Charles W. Schueddig was assigned as the S-4, replacing 1LT Anthony C. Muse, who rotated to CONUS.

- On 6 Jan 70, CPT Richard A. Peace assumed command of the 513th Engineer Company (Dump Truck) from CPT Alan J. Arikian, who rotated.

- On 16 Jan 70, CPT Byren L. Smith, Pipeline Engineer of the S-3 section, assumed command of Co C, replacing CPT Samuel L. Lamey, who was reassigned as the advisor to the ARVN 61st Engineer Battalion.

- On 27 Jan 70, CPT Mark L. Weiss became the S-4, replacing 1LT Charles W. Schueddig, who rotated to CONUS.

- On 30 Jan 70, MAJ Philip J. Galanti became the S-3, replacing MAJ Romayne E. Schroeder who became BN XO.

- CPT Francis J. Lowe succeeded CPT John R. Logue as Adjutant.
- MAJ Vincent J Parmesano, former BN XO, was transferred to the 35th Engineer Group Headquarters.

Attachments

- 513th Engineer Company (Dump Truck)
- 687th Engineer Company (Land Clearing)
- Quarry Platoon, 73rd Engineer Company

Missions and Movements

HHC: The Company performed its normal mission of providing support to Battalion Headquarters. The Utilities Section continued improvements to living conditions in the battalion area by performing light construction and minor repairs. Three water points were operational: Song Pha supported Co C; Song Mao supported 2/1 Armored Cav and 5/27 Artillery; and Phu Quy supported Co D. The water points produced 368,000, 272,000 and 115,000 gallons respectively.

Co A: The Company paved 15.7 kilometers of double-lane roadway on QL-11 and an ammunition haul road for Cam Rahn Bay Support Command. A total of 77,500 cubic yards of rock were crushed during the reporting period. The Direct Support Maintenance Platoon received 260 job orders during the period. The DYNALECTRON civilian contract mechanics continued to provide support maintenance for the MCA-LOC equipment.

During this period the Battalion's support assets for asphaltic paving were gradually absorbed into the Co A. Equipment support to the Battalion from the Company declined sharply during the period because of equipment being reassigned to the Army of the Republic of Vietnam (ARVN) and the redistribution of assets due to shortages.

Co B: The Company completed construction of its portion of QL-11 from Phan Rang to bridge 16, a distance of 26 km. Preparations began for relocation to Vinh Hao, sixty-nine km from Phan Rang on QL-1 at coordinates BN 534472. A fair weather access road was built up a mountainside to a Republic of Korea Army firebase.

Co C: The Company upgraded a seven km section of QL-11 and provided continuous maintenance of two sections of QL-11 totaling 44 km. It constructed a Bailey Bridge construction on QL-1, repaired the Song Mao airfield, upgraded a 1.6 km ammunition area access road in Cam Rahn Bay, and performed miscellaneous maintenance and support tasks at the base camp.

Co D: The Company's primary effort was the upgrading of QL-1. During January, the Company received 30 enlisted trainees from the 63rd ARVN Engineer Battalion to train via OJT (On The Job) on various pieces of equipment and as mechanics. Trainees responded and were becoming efficient operators by close of the reporting period.

513th Engineer Company (Dump Truck): The Company continued its mission of supporting road construction. The Second Platoon, which had been attached to the 864th Engineer Battalion in Nha Trang, rejoined the Company in Phan Rang on 2 Dec 69.

Enemy Actions

- The Phan Rang Air Base was attacked with mortar and rocket fire on 4 Nov, 14 Nov, and 20 Nov. No 589th casualties.
- Three culverts were destroyed by enemy demolitions on QL-11.
- On 21 Nov 69, a Co D work party was ambushed five kilometers south of their base camp on QL-1. The enemy fired thirteen rounds of B-40 rockets and small arms fire. Casualties were light.
- On 22 Nov 69, eleven 82-mm mortar rounds fell short of the Co D base camp at Phu Quy.
- On 3 Dec 69, the last vehicle of a Co C convoy was hit by a Claymore mine on QL-11, near Bridge 11. Casualties were light. *(Author's Note: the ORLL appears to be incorrect in regards to the casualties. 1LT Charles Schueddig, who was in the second vehicle and is now a member of the 589th*

Association, reports that three vehicles were in the convoy and one soldier had serious head and facial injuries and the other soldier lost an eye. 1LT Schueddig also reports that the incident occurred near Bridge 21, not Bridge 11.)

- On 4 Jan 70, six 82-mm mortar rounds fell short of the Co C base camp at Song Pha.

- On 12 Jan 70, the 513th received small arms fire near Bridge 9 on QL-11.

- On 16 Jan 70, a Co C dump truck convoy received enemy small arms and automatic weapons fire. Casualties were light.

- On 16 Jan 70, a Co D dump truck detonated a mine on QL-1. No casualties.

- On 30 Jan 70, a Co D 10-ton tractor/trailer detonated a mine two miles off QL-1, near a borrow pit.

- On 31 Jan 70, a Co D dismounted truck driver detonated an anti-personnel mine near a borrow pit on QL-1. The driver received serious injuries.

1 FEBRUARY 70 – 30 APRIL 70

Personnel

- On 12 Feb 70, CPT Joseph Feast Jr. assumed command of Co B from CPT Daniel S. Quintard Jr., who rotated to CONUS on 16 Feb 70.

- On 13 Feb 70, CPT W. T. Permenter became Battalion Chaplain taking over duties from MAJ Arnold T. Ellsworth, who departed for CONUS on 5 Feb 70.

- On 9 Mar 70, 1LT Elmer C. Hoepker became Pipeline Engineer in the Operations Section.

- On 17 Mar 70, 1LT Russell Bement, a Platoon Leader in Co D, replaced CPT Francis J. Lowe as Battalion Adjutant.

- On 19 Mar 70, 1LT Joseph A. Hauck assumed command of Co A from CPT Robert W. Kyle.

- On 20 Mar 70, CPT Robert W. Kyle became Engineer Equipment Maintenance Officer.

- On 22 Mar 70, CPT Francis J. Lowe assumed command of the 513th Engineer Company (Dump Truck) from CPT Richard A. Peace, who rotated to CONUS on 26 Mar 70.
- On 6 Apr 70, CPT David Darwin, former Civil Engineer, departed the Battalion, reassigned to the 19th Engineer Battalion.
- On 9 Apr 70, 1LT Joseph A. Hauck was promoted to CPT.
- On 9 Apr 70, CPT Edward D. Haggerty became the Civil Engineer of the Operations Section.
- On 18 Apr 70, CPT Ted Bauer became Civilian Personnel Officer.

Attachments

- 513th Engineer Company (Dump Truck)

Missions and Movements

HHC: The Company performed its normal mission of providing support to the battalion headquarters and line companies. The Utilities Section continued to maintain the battalion headquarters by performing light maintenance and minor repairs. All revetments in the Company area were repaired and refilled with sand. Crossbracing was strengthened in bunkers and siding was repaired in revetments around billets. Two water points were in operation: Song Pha supporting Co C, 589th, and Song Mao supporting 2/1 Armored Calvary and 5/27 Artillery. Combined, the units produced 639,000 gallons of water.

Co A: On 4 Feb the Company assumed responsibility from the 554th Civil Engineer Squadron (Red Horse) for operation and maintenance of the Barber Greene Batchpac asphalt plant at Phan Rang Air Base. The entire twenty-six kilometer length of QL-11 from Tan My to Song Pha was capped with a two-inch lift of asphalt during the period. In addition, the Company paved 5.44 km of double lane in the Good View Pass of QL-11. During the report period, 41,000 cubic yards of various sized aggregate were produced. The Direct Support Maintenance Platoon completed 324 job orders of a total 358 submitted, leaving thirty-four open at time of this report.

Co B: During the period the Company moved from Phan Rang to LZ Last Chance (near Vinh Hao) and began operating from the new location. Effort was divided into three main areas—base camp construction, industrial area development, and upgrading highway QL-11. The Company also assumed responsibility for continuous repair of the Song Mao Airfield and completed two Civic Action Projects. The relocation of the asphalt plant and rock crusher to LZ Last Chance was completed during the period.

Co C: During the period the Company was engaged in upgrading the Good View Pass on QL-11, plus an additional thirty-two km of the same highway. The unit was also involved with improving its base camp and reconstruction and maintenance of perimeter defenses. A facility was constructed at Song Pha to permit the unloading of crushed rock from side-dump railroad cars. On 3 April, the Vietnamese National Railway System began hauling rock base course from Phan Rang to Song Pha, doubling to 420 tons per day the amount of base course available to the Company.

Co D: Construction activity during the period included continuation of road upgrading on QL-1, completion of the Song Nao Operational Support Mission, and construction of railroad loading and off-loading facilities.

513th Engineer Company (Dump Truck): During the period the Company provided haul support to the 589th and other units. One mission was to send thirty-five five-ton dump trucks to Hot Rocks on QL-21 near Ban Me Thuot. On 1 Feb 70, the Company had a one-day maintenance stand-down after four difficult days of hauling to the Special Forces Camp at Duc Lap. A total of 271,869 miles were driven by the unit.

Enemy Actions

- The Phan Rang Air Base received attacks—generally small in nature—on 11 Feb, 16 Feb, 13 Mar, 14 Mar, 1 Apr, 5 Apr, 7 Apr and 20 Apr. No 589th casualties.

- On 27 Feb 70, three jeeps of Co C traveling east on QL-11 near bridge 22 encountered a command-detonated mine. No casualties.

- On 28 Feb 70, nine mortar rounds fell short of the Co D base camp.

- On 5 Mar 70, a Co D truck received moderate damage when it backed over a mine. No US casualties, but a Vietnamese child was injured by another mine in the same area.

- On 13 Mar 70, Co D received small arms sniper fire near its base camp's west perimeter.

- On 1 Apr 70, the Co D base camp received 20 rounds of mortar fire. No casualties.

- On 1 Apr 70, the center pier of Bridge 63 on QL-1 was destroyed by enemy explosives.

- On 10 Apr 70, a Co A five-ton tractor received one round of small arms fire through the radiator.

- On 21 Apr 70, a Co D 5,000-gallon water tanker backed over a mine. No casualties.

1 MAY 70 – 31 JULY 70

Personnel

- On 22 May 70, LTC Byron N. Schriever assumed command of the Battalion from LTC Donald A. Ramsay, who departed for CONUS on the same day.

- On 28 May 70, CPT Gerald R. Thiessen assumed command of Co B from CPT Joseph Feast Jr., who left for CONUS.

- On 15 Jun 70, the 585th Engineer Company (Dump Truck) was attached to the 589th, bringing 1LT Thomas L. Barron as Company Commander.

- On 11 Jul 70, CPT Anthony R. Kropp left the position of Assistant S-3 and assumed command of Co C from CPT Byron L. Smith.

- On 12 Jul 70, 1LT Donald E. Alsted, the Battalion Signal Officer, departed. His duties were assumed by the Adjutant, 1LT Russell Bement III.

- On 15 Jul 70, 1LT Larry W. Owen assumed command of Co A from CPT Joseph A. Houch, who moved to the 35th Engineer Group to manage the Industrial Complex Center.

- On 17 Jul 70, CPT Mark L. Weiss, S-4 Officer, departed for CONUS.
- On 18 Jul 70, CPT Byron L. Smith was assigned as S-3 Officer, replacing MAJ Philip J. Galanti, who departed for CONUS.
- On 22 Jul 70, 2LT Edmond Brown Jr was assigned as S-4 Officer.

Attachments

- 513th Engineer Company (Dump Truck)
- 585th Engineer Company (Dump Truck)

Missions and Movements

HHC: The Company provided its normal support to Battalion Headquarters and the line companies. The Utilities Section improved living conditions in the HHC area by replacing existing revetments with stronger ones. The water points at Song Pha in support of Co C, and at Song Mao to support part of Co B and 2/1 Calvary, produced 673,000 gallons of water.

Co A: The Company's industrial complex was expanded by the addition of a Pioneer Model 300 roll crusher unit and a Cedar Rapids MCA soil stabilization plant. Paving operations for the period included 5.45 km of double lane asphalt on QL-11 in Good View Pass and 9.36 km of double-lane on QL-1. An asphaltic soil stabilization plant was erected at the industrial complex site and placed in testing operation. During the period, 56,747 cubic yards of various sized rock was produced. The Direct Support Maintenance Platoon completed 223 of a total 284 job orders submitted during the reporting period. The increase of open job orders—as compared to thirty-four at the end of last period—is due to a large number of parts requisitions still in the "due out" category. These parts included critical items such as engines, transmissions and transfer cases.

Co B: The Company continued primary effort in completion of its industrial complex set up and final grading and organization of its base camp. Due to a change in TOE and the formation of a 31-man provisional asphalt platoon, Co B's personnel strength increased from 115 men on 1 May to 173 men on 15 July

1970. An operational support project, serving the 5/27 Artillery Battalion, was undertaken for the construction of Fire Base Mike Norton atop a granite hill.

Co C: The Company was engaged in upgrading the Good View Pass on QL-11 and maintaining traffic on another 31 km of the same road. Work was also done in the cantonment area to improve the perimeter defense and ammo storage.

Co D: Primary effort was placed on upgrading QL-1, constructing an additional offloading facility for rail haul of base course from Phan Rang, constructing culverts and bridges on QL-1, and repairing culverts on QL-11 damaged by enemy activity. During the period the Company trained and graduated nine ARVN Engineer Soldiers.

513th Engineer Company (Dump Truck): The Company supported the 589th and the 577th Engineer Battalions.

585th Engineer Company (Dump Truck): The Company supported the 20th, 589th, and 577th Engineer Battalions. On 15 June the unit was attached to the 35th Engineer Group (Construction).

Enemy Actions

- Phan Rang Air Force Base received harassing mortar and rocket attacks on 6 Jun, 10 Jun, 2 Jul, 10 Jul, and 21Jul. No 589th casualties.
- On 2 Jul 70, an A Co ten-ton truck was hit by shrapnel near the Air Force Quarry.
- On 23 May 70, a Co D 290M tractor ran over a mine, with heavy damage to the tractor, but no casualties.
- On 23 May 70, a Co D 5-ton dump truck hit a mine while backing into a sand pit area. Light vehicle damage, no casualties.
- On 30 May 70, culvert QL-11-15 east of Tan My was 100 % destroyed.
- On 2 Jun 70, a railroad bridge near QL-11-12 was blown. The damage temporarily interfered with the rail haul of base course to Co C.
- On 12 Jun 70, culverts QL-11-17 and QL-11-16.10 were damaged.

- On 29 Jun 70, culvert QL-11-12 was completely destroyed.

- On 6 Jul 70, two meters of railroad track between Phan Rang and Co C were damaged by explosives.

- On 17 Jul 70, culverts QL-1-43 and QL-1-45 received enemy damage.

1 AUGUST 70 – 31 OCTOBER 70

Personnel

- CPT Francis J. Lowe was Commander of the 513th Engineer Company (Dump Truck) from the start of this period until 23 Aug when 1LT Michael T. Heidt assumed command. 1LT Robert N. Burnham assumed command on 21 Oct.

- On 9 Aug 70, CPT Terence C. Holland assumed command of Co A from 1LT Larry W. Owen who became Platoon Leader of the Asphalt Platoon.

- On 4 Sep 70, 1SG Harry W. Long arrived to replace 1SG Keown who rotated to CONUS on 7 Sept.

- On 17 Sep 70, MAJ Byrnes, as XO, was assigned to be a Task Force Commander of the different elements located at Vinh Hao.

- On 21 Sep 70, 1LT Ruszczyk took over the Asphalt Platoon from 1LT Owen, who transferred to Co B.

- On 3 Oct 70, 1LT Riley was loaned to Co A from HHC, to run the ASL Section on the departure of CW2 Brannon.

- On 6 Oct 70, CW2 Brannon moved to Co B.

- On 18 Oct 70, SFC Franco arrived to become Platoon Sergeant of the Quarry and Equipment Platoon.

- On 13 Oct 70, 1LT Lee Man Kyou, Corps of Engineers Republic of Korea Army and six enlisted men joined Co A for training on engineer equipment.

- CPT Thiessen was reassigned to the 35th Engineer Group staff.

- 1LT Hultgren and 1LT Strub were assigned to the 585th Engineer Company (Dump Truck).

- 1LT Payant, 1LT Owen, 1LT Jackson and 1LT Twitty were the officer gains for Co B.
- 1LT Ashmore was moved from 2nd Platoon Leader to Construction Officer of Co B.

Attachments

- 513th Engineer Company (Dump Truck)
- 585th Engineer Company (Dump Truck)
- Earthmoving Platoons from the 577th Engineer Battalion
- Earthmoving Platoons from the 84th Engineer Battalions

Missions and Movements

HHC: The Company performed normal functions during the period. Special emphasis was placed on motor maintenance in preparation for CMMI inspections. The Utilities Section completed a building for use as the Battalion Dental Clinic. Other carpenters were sent to Co B to provide support of facilities at Vinh Hao. The two water points produced 177,289 gallons of water in support of Co C, 2/1 Cav, MACV, and 5/22 Artillery.

Co A: The Company remained at Phan Rang Air Base and continued to work on assigned missions of producing rock and asphalt and providing Direct Support Maintenance for the Battalion. During the period, 84,163 CY of various size aggregate and 24,376 tons of asphalt were produced. An extra platoon, the Asphalt Platoon, was organized during this reporting period. It consists of the asphalt plant, the stabilization plant and the paving train. Although not authorized by the TOE, this platoon has proven to be useful to provide control over the elements which constitute it. The Asphalt Platoon produced a total of 24,376 tons of asphalt and stabilized base during the period. During the period the Direct Support Maintenance Shop processed 505 job orders in support of the battalion and attached units.

Co B: The unit had a tremendous change in personnel strength with the arrival of three Earth Moving Platoons from the 577th and 84th Engineer Battalions.

Construction continued on QL-1 with 214,961 cubic yards of fill and base course placed. With the arrival of personnel, considerable alterations and additions had to be made to the base camp. Maintenance at the Song Mao Airfield continued.

Co C: Work continued on road repairs at the Good View Pass and on maintaining traffic on RVN Route QL-11. This work consisted of repairing pavement damage, reconstructing ditches and extending shoulders. The Company also engaged in improving the base camp and maintaining perimeter defenses.

Co D: Primary emphasis was placed on upgrading and on culvert construction on QL-1. This work has been progressing at an accelerated rate due to the addition of four Earthmoving Platoons to the Co D horizontal efforts. Presently attached are the Earthmoving Platoon from Co C and all Earthmoving Platoons from the 577th Engineer Battalion.

513th Engineer Company (Dump Truck): The primary mission was support of the 589th with dump trucks. Company Headquarters and the First Platoon were stationed at Phan Rang; one squad of the First Platoon was stationed at Vinh Hao in support of Co B; the Second Platoon was stationed at the top of Good View Pass, attached to Co A, 577th Engineer Battalion.

585th Engineer Company (Dump Truck): During the period the Company supported the 589th and 19th Engineer Battalions.

<u>**Enemy Actions**</u>

- On 3 Aug 70, two Co D personnel were wounded by a mortar round that landed in its base camp.
- On 15 Aug 70, the CPO vehicle received one round of 60-mm fire on QL-11. No casualties or vehicle damage.
- On 16 Aug 70, at BN 527443 Bridge #32, a concrete bridge span was rigged with three enemy electrical charges. One charge went off; the other two were removed by Co B.
- On 18 Aug 70, at BN 624522, the 585th received one round to the right of the lead truck, and 4 to 6 rounds towards the gun truck. No casualties.

- On 31 Aug 70, at BN 599992, Co C vehicles spotted a VC flag booby trapped with a 60-mm round lying under the flag on the road. The NCO in charge gave the order to stay clear, but one enlisted soldier ignored the order and reached for the flag. He detonated the booby trap and was killed. As the Medevac helicopter came in, the pressure from the blades set off another mine, with no casualties.

- On 1 Sep 70, at BN 598988, a convoy en route from Co C to Phan Rang received small arms fire and automatic weapons fire. No casualties, light vehicle damage.

- On 21 Sep 70, a portion of the railroad tracks between Phan Rang and Song Pha were blown.

- On 25 Sep 70, at BN 543505, an 830M scraper bobtail of the 577th hit a mine. No casualties, light vehicle damage.

- On 29 Sep 70, at BN 523610, a Co B scraper's rear tire was damaged by a command-detonated mine.

- On 30 Sep 70, the Co B soils people on QL-1 received six rounds of automatic weapons fire. No casualties.

- On 6 Oct 70, the VC blew out a 36-inch culvert in the Good View Pass, Co C's area of operation.

- On 15 Oct 70, at BN 539471, a 5-ton dump from the 513th ran over a mine. No casualties, heavy vehicle damage.

1 NOVEMBER 70 – 31 JANUARY 71

No ORLL available.

1 FEBRUARY 71 – 30 APRIL 71

Personnel: No personnel changes were listed.
Battalion Commander LTC Donald M. O'Shei signed this ORLL.

It was learned in January that the 589th Engineer Battalion had been selected as one of the stand-down units. Planning was started to complete the assigned projects before the stand-down date. The Battalion was not to be deactivated but to be drawn down to 49 spaces. These spaces were required to staff the ASL and third shop of Co A which remained to support the Vinh Hao Industrial Site and the 299th Engineer Battalion (Combat). The schedule for the drawdown was:

UNIT	CEASE WORK	DRAWDOWN COMPLETED
C/589th	15 Mar 71	29 Mar 71
B/589th	21 Mar 71	5 Apr 71
D/589th	28 Mar 71	12 Apr 71
A/589th	4 Apr 71	19 Apr 71
HHC/589th	11 Apr 71	26 Apr 71

(Author's Note: CSM Moore once informed the 589th Association that whether the 589th or the 577th would draw down was decided by a coin toss.)

MISSIONS AND MOVEMENTS

HHC: The Company continued its normal function of providing support to the Battalion Headquarters during the reporting period. The Utilities Section continued to improve living conditions in the battalion area and at Vinh Hoa by performing light construction and minor repairs. The water point at Song Pha produced 88,000 gallons of water in support of Co C before it was shut down on 1 Feb and returned to Phan Rang for transfer. The water point at Song Mao produced 89,200 gallons before it was laterally transferred to the 864th Engineer Battalion (Construction).

Co A: The Company continued its assigned missions of producing rock and asphalt and providing direct support maintenance to the Battalion. The Company remained located on Phan Rang Air Base and retained its modified organization of four platoons (Headquarters, Quarry and Equipment, Asphalt, and Maintenance). During the period a total of 138,365 cubic yards of various sized aggregate were produced. Asphalt production increased during this period due to the increase in prepared base course. The maintenance shop processed 468 job orders of which 374 were closed out. Thirty-one of these were for engine changes in 5 Ton Dump Trucks.

Co B: The Company and attached units placed most of their effort on completion of the road program. These projects included 7.6 kilometers of earth fill (146,106 cubic yards), 13.1 kilometers of crushed base course (39,399 cubic yards), 15.8 kilometers of asphaltic concrete paving (17,155 cubic yards), and 21.4 kilometers of road shoulder stabilization.

Co C: During this period the Company engaged in upgrading QL-11 on Good View Pass, redecking Bridge 16 and improving the Song Pha base camp and perimeter defense.

Co D: The primary effort during this period was installing culverts and upgrading QL-1. Vertical work included placing: four single-barrel twenty-four-inch culvert sites, three single-barrel, and three triple barrel thirty-six-inch culvert sites; four single-, two double- and two triple-barrel forty-eight-inch sites; two single and three double barrel sixty-inch sites, and one single seventy-two-inch culvert site. Upon completion of QL-1 the Company was tasked with construction of a maintenance pad and helicopter hover lanes for A/7/17ACS, which included preparation of the site for paving and construction of drainage ditches and structures.

513th Engineer Company (Dump Truck): During this period the Company supported the 589th and 577th Engineer Battalions with dump truck haul capacity. A program to train Vietnamese Nationals to drive MCA dump trucks was initiated on 7 Dec 70 with the hiring of twenty-five Vietnamese. After a two-week training period twenty-three were licensed and took over the operation of MCA trucks. On 16 Apr 71 the entire 513th Engineer Company (Dump Truck) departed Phan Rang for relocation with the 864th Engineer Battalion (Construction).

585th Engineer Company (Dump Truck): The Company continued its mission to provide dump truck haul support to the 589th. With the coming of the 589th drawdown, Company Headquarters and one platoon moved to the Vinh Hoa Industrial Site, where they became the backbone of Task Force Asphalt. In conjunction with this move, one platoon was sent to Dillard Industrial Site to provide dump truck support to the 815th Engineer Battalion (Construction).

1 MAY 71 – 31 JULY 71

No ORLL available for this period.

1 AUGUST 71 – 31 OCTOBER 71

No ORLL available for this period.

(Author's Note: The remaining 589th members were transferred to other engineer units within Vietnam. The remaining Co B was assigned to the 864th Engineer Battalion (Construction), and the remaining Co C was assigned to the 577th Engineer Battalion (Construction). CPT Frank Lowe, the former 589th Adjutant, returned the 589th Battalion flag to the United States.)

PART 3

OPERATIONS, ORGANIZATION AND LOGISTICS
ISSUES AND LESSONS

Background

Construction battalions went to war in Vietnam with TOEs designed for the conventional warfare of World War II and the Korean War. This issue is described in greater detail in the above section: "Historical Background." Yet, as reported in the ORLL *Enemy Actions* sections above, the 589th was engaged in insurgency warfare and often operated over long distances. This situation created a need for more heavy automatic weapons, more indirect fire weapons (i.e. grenade launchers), more radios, and additional associated training for the additional equipment. More soldiers needed to be trained in calling for Medevac, artillery fire, and air support. There were also requirements for specialized engineer equipment for working in the varied Vietnam terrain.

The U.S. Army had a system for modifying TOEs, resulting in Modified TOEs, called MTOEs. Unfortunately, the review process for modifying TOEs was slow and cumbersome, and did not meet unit requirements for rapidly changing missions. Two of the Battalion Commanders recommended that a pool of equipment and specialists be established at a higher headquarters that field commanders could use for obtaining additional equipment and specialists on a temporary basis without going through the entire time consuming MTOE approval process. The authors have not been able to learn if such a pool of equipment and specialists was ever established.

This was very much a "logistical war." Construction materials had to be moved from the United States to specific 589th job sites in a timely manner. Many of the issues and lessons listed below describe this logistical war.

The following is a selection of the operational, organizational, and logistics issues and lessons extracted from the 589th ORLLs that offer a flavor of construction battalion operations in Vietnam. Much greater detail is provided in the ORLLs themselves. The full ORLLs are available in the *History* section of the 589th Engineer Battalion Association (Vietnam) website, http://www.589thEngineers.com, and are available for use by the public.

1. Deploying units should insist on sending an escort party with their equipment ship. The escort party can greatly assist in assuring that equipment is carefully loaded and unloaded. Minor maintenance can be immediately performed to process vehicles upon unloading. This party should consist of either the Maintenance Warrant Officer or Maintenance NCO and several mechanics, with toolboxes.

2. The guerilla-type warfare conducted by the Viet Cong required that constant security be kept simultaneously at cantonment camps and at engineer work sites. These combined security requirements often resulted in fifty percent of available engineers being used as security forces, while only fifty percent of engineers were actually doing engineer work. This greatly slowed the engineer effort (which was the enemy's intent). Occasionally, this situation was alleviated through the use of Vietnamese and Korean soldiers as security forces.

3. The increased security needs meant more engineer soldiers had to be trained on machine guns and grenade launchers in order to simultaneously maintain job site security during the days, and twenty-four-hour per day security in the cantonment camps.

4. Construction battalions operating in Vietnam often operated across greater distances and at more sites than did construction battalions in previous wars. Yet, the "conventional war" TOE of construction battalions operating in Vietnam did not provide sufficient radios and heavier automatic weapons needed for the enemy engagements, distance, and multiple work site situations encountered by the construction battalions in Vietnam.

5. Construction battalions operating in Vietnam often converted 2 ½ ton trucks to "gun trucks," mounted with machine guns for use in convoy and work site

protection. These conversions reduced the haul capabilities of the construction battalions.

6. Mine warfare was the enemy's favorite weapon against construction battalions. The enemy's use of mines created a number of operational and related organizational issues:

   a. Roads were often cleared every morning by combat engineers, supported by infantry units providing security. These minesweeping operations often did not, however, cover engineer work sites such as landfills, culverts, and bridge banks. Thus, construction battalions had a need for more minesweeping equipment and more trained mine-sweeping operators to sweep for mines in construction sites.

   b. Passengers in small vehicles such as ¼ ton jeeps and ¾ ton trucks were particularly vulnerable to mine detonations. In many areas, vehicles of this size were not allowed on the roads.

   c. Grader operators were particularly vulnerable to mine detonations.

   d. Dozer "shotguns" riding in the dozer right side jump seats were also vulnerable to mine blasts that occurred under the right side dozer tracks.

   e. Placing asphalt on roads greatly reduced mine incidents because such mine placements were much easier to detect. In response, the enemy placed more mines on road shoulders and in areas where tractor-trailers often went off the asphalt to make wide turns. As a result, it was important to place asphalt in these areas as well.

   f. The Battalion experimented in hasty mine clearing on Route TL-3A by rolling the road with a sheepsfoot soil compaction roller and dozer. It was learned that the spacing between the sheepsfoot's rollers did not assure detonation of all pressure-sensitive mines in one pass. Thus, two or more passes were needed to assure that all mines were detonated. (Note: higher headquarters was not happy with this experiment.)

7. The B/589th work site at Di Linh was attacked on the night of 15 Sep 68 by small arms fire. The platoon leader decided to keep the two M-60 machine guns mobile so that maximum fire-power could be brought to bear on the areas where enemy fire was heaviest. Both of the machine guns were moved to several locations during the fight. The next day, during interrogation of suspects, it was learned that the enemy believed that there were thirteen machine guns in the camp.

8. Support units (e.g. dump truck, power distribution, water well drilling, asphalt, concrete mixing and placing, land clearing, etc.) were frequently required to give support to two or more units in separate locations. This often created inadequacies in both maintenance and personnel support. These units should have their TOEs and Standard Operating Procedures (SOPs) revised to enable them to operate in multiple locations with adequate maintenance and personnel support at each location.

9. The TOE organization of the Battalion and company supply sections proved entirely inadequate for the management of construction materials simultaneously with the administration, storage, and issue of rations, clothing, and general supplies. The Battalion as a whole managed 100 to 150 tons of supplies and materials per day.

10. The Battalion S-4 section was not authorized forklifts to handle construction materials. This non-availability of forklifts resulted in the inefficient use of manpower and construction equipment to offload and move construction materials. Each construction battalion needed at least two 10,000-pound capacity forklifts when the companies were nearby, and one additional 10,000-pound fork lift for each company that was physically separated from its battalion.

11. Detailed packing lists were absolutely essential while unloading in-country. Copies of each packing list should have been placed both outside and inside each container, and copies should have been provided to a number of individuals so as to prevent loss of lists. A unit master listing of package lists is essential.

12. Project designers should have used only materials from a command-wide approved construction materials stockage list. Materials not in command stockage took too long to procure.

13. The materials transportation system required an average of eleven days reaction time upon notification of items to be shipped from the depot to the 589th. In response, the 589th often used its own trucks to pick up materials from depot, thus reducing the availability of trucks for use on job sites.

14. The Battalion did not have sufficient transits and levels to meet survey requirements. Additional authorized transits and levels would have enabled the survey teams to be split, with non-surveyors being used to expedite survey work.

15. The many projects in which the Battalion was involved required the work of surveyors. However, surveying capability was limited by TOE, and project progress was slowed by lack of surveying capability. The authorization of three Abbey Hand Levels per construction platoon would have enhanced the capability of platoons to perform the minimum of survey required to maintain progress.

16. The Battalion relied heavily on Vietnamese indigenous labor for vertical construction assistance. The use of this labor was limited, however, because the Battalion was not authorized additional tool sets for use by this added labor force.

17. Loose cement stored in hoppers for more than two or three days in the humid Vietnamese environment tended to harden and clog in hopper doors and chutes.

18. Concrete, when placed during periods of high temperature and/or winds, tended to cure too rapidly. As a result, finishing was difficult and cracking occurred frequently. In addition, inexperienced personnel tended to add water to the mix to make placement easier, thereby changing the cement-to-water ratio and weakening the concrete. Rapid curing was overcome by

placing the concrete early or late in the day, providing shade, and first soaking the ground to prevent water movement from the concrete into the ground.

19. Where culvert headwalls on temporary bypasses were likely to wash out during monsoon rains, a sand-cement mix was placed in sandbags and reinforcing rods were driven though them vertically to improve stability.

20. When a roadway had to be constructed through a rice paddy, it was found that removing the muck to get to more stable ground only resulted in hopeless, repeated bogging of heavy equipment. The 589th learned that the most expeditious method of providing a stable base was to fill the area with a mixture of rock and laterite soil to a depth, in most cases, of three feet. This provided a stable enough sub-base for heavy compaction equipment and subsequent base course fill.

21. Upgrading road shoulders with base course and applying double surface treatment required equipment critically needed for other road building tasks, consumed much time and effort, and resulted in a non-durable product. A better solution was that of placing one-and-one-half inches of asphalt on the shoulders.

22. Even in areas of low enemy activity, running dump trucks by themselves often led to cases of speeding problems and serious accidents due to carelessness. Small convoys of six to eight trucks afforded both protection and maximum utilization of haul capability by decreasing loading and unloading times and decreasing the number of control and protection vehicles. This was not a hard-and-fast rule, since small convoys could not properly be utilized in areas where traffic control was a major factor, such as mountain passes or built-up areas. In these instances, larger convoys were more practical, since all trucks could clear the congested area at one time.

23. The noise of twenty-five to thirty dozers operating simultaneously during land clearing made it extremely difficult to quickly discover a dozer that was having problems. The use of smoke grenades proved effective in signaling when problems developed.

24. When "floating" Rome Plow blades in sandy soil, much brush was merely pushed over and not cut. In these conditions, operators should pitch the Rome Plow blade forward and attempt to keep the blade three to four inches underground.

25. Totally clearing bamboo clumps proved to be extremely difficult. Rome Plows and bull blades often just ripped the stalks without cutting them. By raising either type of blade up to near maximum elevation and pushing the clump partially over, the edge of the root mass was lifted. Then, by backing and digging, this root mass was easily dug up and the entire clump destroyed.

26. While clearing in known VC assembly areas, observation towers were often found in many large trees. These towers could be located by their observing long pieces of bamboo with branches partially broken off and attached to the trees in the manner of a ladder.

27. Training films on venereal disease, drug abuse, and first aid were difficult to impossible to obtain. Visual training aids such as films would have greatly assisted in informing troops on the adverse effects of venereal disease and drug abuse.

28. The construction battalion TOE was inefficient when the predominant Battalion mission was that of road and airfield construction. In this situation, the Battalion operated more efficiently in a functional organizational structure under which squads, platoons, and companies became specialists in one function such as rock blasting, rock production, earthwork, sub-base, base, paving, hauling, etc.

29. The scope and magnitude of Battalion operations grew to the point at which the full-time assignment of a helicopter would have greatly facilitated mission accomplishments. Battalion road maintenance responsibilities eventually comprised 170 km of highway. In addition to road construction and maintenance, operational support tasks (e.g. base camp, artillery site, and airfield construction) were frequently required at points as much as 165 km apart. An aircraft would have insured tighter command and control; reduced

medical evacuation reaction time; and would have immeasurably enhanced perpetual reconnaissance activities.

30. The following is an extract of the 31 January 69 ORLL reflecting the Battalion Commander's continued frustration with the system for modifying TOEs, and the difficulties in getting supplies and equipment:

"The object of this discourse has been to present a "Grass Roots" view of the logistical picture; to present facts and cite instances which might provide clues to spotting flaws in the system. No TOE can be written to satisfy all requirements generated by changing conditions. The nomadic nature of the engineers, specific mission, location, enemy situation and weather all affect the accomplishment of the mission. The degree to which mission accomplishment is affected is dependent upon whether the right tool or right item of equipment is on hand for the right job. Engineer Construction Battalions are being tasked to provide work of a sophistication and nature not envisaged by the TOE, and the system is not responsive enough to satisfy a valid demand for non–TOE item in the time frame required. A field commander should be allowed to compare his TOE against his mission, order, and expect the equipment to arrive within a reasonable time frame thereafter. This would entail building a theater float of certain selected items of tools and equipment and increased quantities of "expected expanded use" items, i.e. radios, weapons, etc. More maintenance people and repair parts would also be required. The equipment would be returned to float as soon as possible as the job was completed or when the conditions which justified their acquisition cease to exist."

31. By 1970 the Battalion was involved in two separate programs to reduce the size of American forces in Vietnam and to increase the participation of local nationals in the LOC (i.e. road and airfield construction) program by augmenting U.S. units with civilian employees. The two programs were implemented separately and in piecemeal fashion with no coordination between them to integrate the loss of U.S. personnel with the concurrent addition of local nationals. The timely phase-in of qualified Vietnamese

replacements did not occur because of the total absence of any apparent coordination between military and civilian personnel organizations.

32. Handling the administrative details of a local national work force required under a Type B TOE organization required an officer and staff on a full-time basis for the administration of personnel records and pay for members of the local national work force.

33. One can see from the ORLL Personnel section that officers did not remain in their positions for long. Most enlisted soldiers served in a single position for only one year due to their being assigned to Vietnam for one year. This continuous turnover of soldiers was a constant operational problem in Vietnam. The Army converted to deploying entire units (as opposed to individual assignments) starting with the Gulf War because of the lessons learned from Vietnam regarding individual rotational assignments.

PART 4
AWARDS AND CITATIONS

1. **The 589th Engineer Battalion (Construction) and assigned units: HHC, Co A, Co C and Co D were awarded the Meritorious Unit Commendation, Department of the Army General Order 43, 68.**

"For exceptionally meritorious achievement in the performance of outstanding service: The 589th Engineer Battalion (Construction) distinguished itself in support of military operations in the Republic of Vietnam during the period 29 April 1967 to 31 December 1967. Within days after arriving in-country, the battalion began to fulfill its counterinsurgency mission by upgrading the beach and access road at the port of Qui Nhon. The four-hundred bed evacuation hospital the battalion constructed for the Republic of Korea's Tiger Division provided excellent medical facilities for this staunch ally and significantly enhanced relations between the United States and Korean personnel. The unifying thread of the unit's efforts was QL-19. Working in conjunction with a civilian contractor, the 589th Engineer Battalion (Construction) transformed this rough, primitive road into a modern two-lane highway. Completion of this important project resulted in a substantial increase in the westward flow of supplies to the critical areas of Pleiku, Kon Tum, and Dak To, while simultaneously decreasing wear and tear on the vehicles utilizing the route. The most significant operational support mission undertaken by the battalion was the construction of a type II light-lift airfield at Vinh Thanh. Characterized by the quality of the work and its remarkable durability, this facility is one of the best double bituminous surface treated airfields in Vietnam. Continuously fostering friendly relations with the Vietnamese people, the battalion built a two story high school in Binh Khe, assisted in the construction of the Go Boi Bridge over the Song Am Phu River and furnished a grader to maintain streets and drainage in Phu Phong (Binh Khe). The remarkable proficiency and devotion to duty displayed by the members of the 589th Engineer Battalion (Construction) are in keeping with the highest traditions of the military service

and reflect distinct credit upon themselves and the Armed Forces of the United States."

2. **B Co, 589th Engineer Battalion was nominated by the 18th Engineer Brigade for the 1970 Itschner Award for mission related accomplishments during 1969.**

   This award is presented by the Society of American Military Engineers to the most outstanding U. S. Army engineering company during a year.

   B Company's primary mission during 1969 was to upgrade route QL-11 in Ninh Thuan Province from Phan Rang to the village of Tan My, a distance of 26 kilometers. This project included construction of four major concrete and steel bridges, emplacement of 29 drainage structures and one major excavation. The Company's missions also included maintenance of a 107-mile section of QL-11 through Ninh Thuan Province to insure it remained open to U.S. and ARVN convoys.

   In his nomination letter, the 18th Engineer Brigade Commander, Brigadier General J. W. Morris stated, *"Of the 73 companies assigned to this Brigade, B Company has significantly impressed me as epitomizing the fulfillment of our Engineer Mission. Whether LOC, Base Construction, or Operational Support, this unit has earned the respect of those units familiar with their accomplishments. I have no doubt but that B Company, 589th Engineer Battalion is the finest engineer company in the United States Army, and that it is fully deserving of the 1970 Itschner Award."*

3. **The 589th was cited in *The History of the 18th Engineer Brigade* for its work on QL-11, especially the work done rehabilitating QL-11 in the Good View Pass. The *History* states:**

   "...the most difficult stretch of roadway that the Brigade had ever undertaken, the 27 kilometer stretch of National Highway in the Central Highlands known as the Goodview Pass, was completed. This road was transformed from a treacherous mountain path into a thoroughfare worthy of the name National

Highway. The Good View Pass metamorphosis is one of the high points of the Lines of Communications highway project that the Brigade is engaged in."

PART 5
"THOSE WE LOST"

A Toast:

To the Brothers we lost in Vietnam;

To the Brothers we have lost since;

Who, while they do not remain with us in body;

They remain with us in spirit.

To Our Absent Brothers

589th Vietnam Losses					
Last Name	Given Names	Command	CO.	Date of Casualty	State of Record
BOGGS	DONNIE REX	585th DT	C	26-Apr-1970	MO
CLAYBORN	BILLY JOE	589th	C	3-Dec-68	AR
CLIFTON	WILLIAM A	589th, 51st	A	26-Jun-1968	AR
DELANO	DARWIN JAMES	589th	B	26-Nov-1968	NH
ENMON	DAVID JERRELL	589th	D	15-Jun-1967	TX
GOODMAN	NORMAN ELAN	589th	C	14-Jan-1969	MI
GRAY	RAYMOND HENRY	589th	B	29-Nov-1970	D of C
HASTINGS	CARLETON PHILIP	589th	B	18-Jun-1968	NY
KINGREY	EDWARD LEO	589th	C	14-Jul-1969	KY
MEEHAN	DONALD LLOYD JR	589th	C	7-Feb-1971	IL
MEEKER	EDWARD HOWARD JR	589th	B	18-Jun-1968	NJ
MOE	RONALD JOHN	589th	B	26-Nov-1968	MT
MORRIS	EDWARD	589th	D	16-Jun-1967	OK
RED HAWK	JESSE MILTON	589th	A	10-Nov-1968	SD
ROPETER	LESTER EARL	589th	D	19-Sep-1968	NY
SCHMIDT	LAWRENCE EDWARD	589th	A	4-Nov-1969	WI
SHERLIN	FREDDIE MICHAEL	589th	D	8-Aug-1969	TN
SNOW	MILTON	589th	B	11-Oct-1967	NY
SPARKS	RICHARD L	589th	D	19-Sep-1968	OH
THOMAS	HENRY BENNY	589th	C	31-Aug-1970	PA
THOMPSON	GREGORY CARL	589th	A	26-Jul-1970	WA
TYLER	JESSIE JAMES	589th	D	8-Aug-1969	SC
WERNER	ANTHONY ROBERT	589th, 51st	A	8-May-1969	OH

Vietnam Losses – Once Assigned to the 589th					
Last Name	Given Names	Command	CO.	Date of Casualty	State of Record
DUTY	ANTHONY	589th & 52nd Inf, 198th LIB	A	7-Jun-1969	KY
McNABB	JOHN JOSEPH	589th & 73rd	A	30-Nov-1967	MA
PENA	DANIEL JR	589th & 84th	D	3-Feb-1968	TX

Information regarding the circumstances of the loss of each 589th member is available from the National Archive Records. This information is also available on the 589th Association web site at: http://www.589thEngineers.com/vietnam-losses.html.

Part 5: *Those We Lost*

PART 6

PHOTOGRAPHIC HISTORY

The following is a selection of photographs that provide a representative sampling of the 589th's area of operations and missions. The "VN Photos" section of the 589th Engineer Battalion Association (Vietnam) website has a much greater selection of photographs. This website is available to the public at "http://www.589thEngineers.com".

Figure 1
Deployment: Preparing Vehicles for Movement @ Fort Hood (March 1967)
Photo by Lawrence Doff

Figure 2
Deployment: Rail Loading for Movement @ Fort Hood (April 1967)
Photo by Lawrence Doff

Figure 3
Deployment: Ship Heading for Vietnam (April 1967)
Photo by James McCarthy

Figure 4
Deployment: Ship Heading Under Golden Gate (April 1967)
Photo by Bill Carter

Figure 5
Deployment: At Sea (April 1967)
Photo by Dave Harbach

Figure 6
Deployment: Aboard Ship
Photo by Mike Wilder

Figure 7
Redeployments Within Vietnam Were Often Done By Sea
Photo by Keith Swilik

Figure 8
Replacements Came Thru Replacement Companies in Long Binh and Cam Ranh
Photo by Dennis Cluth

Part 6: Photographic History

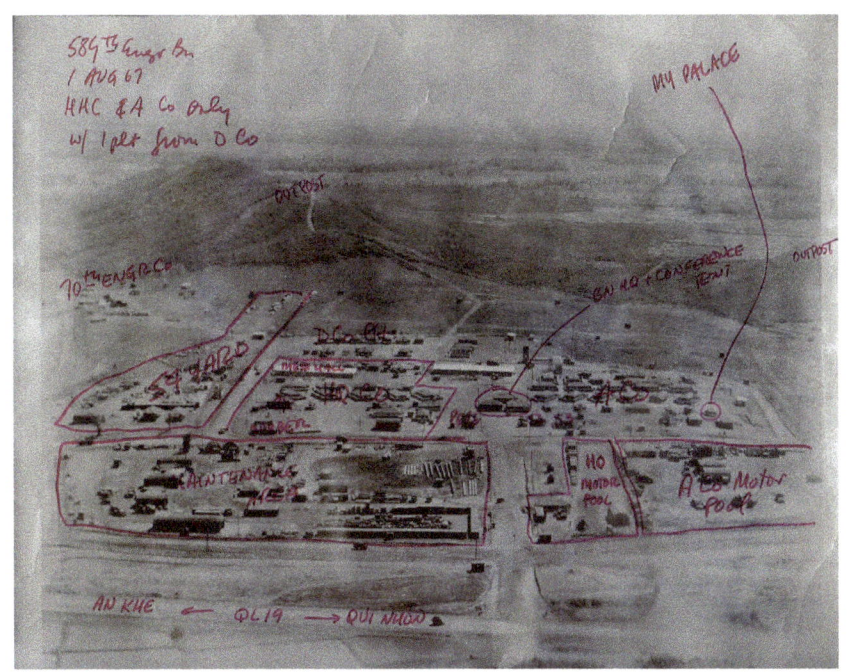

Figure 9
Base Camps: 589th Headquarters / Co A @ Cu Lam Nam (May 1967 – July 1968)
Photo by Lawrence Doff

Figure 10
Base Camps: 589th Headquarters / Co A @ Cu Lam Nam (May 1967 – July 1968)
Photo by Lawrence Doff

Figure 11
Base Camps: 589th Headquarters @ Phan Rang Air Force Base (August 1968 – April 1971)
Photo by Jim Swick

Figure 12
Base Camps: 589th Headquarters / Co A / Co B @ Phan Rang Air Force Base During Monsoon Rains
Photo by Fred Osterman

Figure 13
Base Camps: Cha Rang – Company C (May 1967 – July 1968)
Photo by Keith Swilik

Figure 14
Base Camps: Camp Evans – Earthmoving Platoon, Company C (April – August, 1968)
Photo by Keith Swilik

Figure 15
Base Camps: Wunder Beach – Earthmoving Platoon, Co C (March – August 1968)
Photo by Keith Swilik

Figure 16
Base Camps: LZ Jane – Earthmoving Platoon, Company C (April – August 1968)
Photo by Keith Swilik

Figure 17
Base Camps: LZ Jane – EM Platoon, Company C (See Brown Brush – Agent Orange)
Photo by Dennis Cluth

Figure 18
Base Camps: Don Duong – Company D (September 1968 – March 1969)
Photo by Robert Burden

Figure 19
Base Camps: Phu Quy – Company D (April 1969 – April 1971)
Photo by Frank Lowe

Figure 20
Base Camps: Vinh Hoa – Company B (March 1970 – April 1971)
Photo by Don Ramsay

Figure 21
Base Camps: Song Pha Billet Area – Company C (Aug 68 – Apr 71) (Note Blast Walls)
Photo by Larry Jinkins

Figure 22
Base Camps: Song Pha Motor Pool – Company C (August 1968 – April 1971)
Photo by Bill Stafford

Figure 23
Base Camps: Song Pha – Company C Today
Photo obtained by Larry Jinkins

Figure 24
Tactical: Weapons Practice
Photo by John Van Zelf

Figure 25
Tactical: Bunker and Tower
Photo by Chuck Schueddig

Figure 26
Tactical: Blast Walls @ Phan Rang
Photo by Fred Osterman

Figure 27
Tactical: Base Camp Perimeter Defensive Devices
Photo by Lawrence Doff

Figure 28
Tactical: Gun Truck
Photo by Royce Lloyd

Figure 29
Tactical: Protection Provided by Vietnamese Soldiers
Photo by Dave Harbach

Figure 30
Tactical: Typical Convoy Thru a Congested Town (Song Pha)
Photo by Bill Stafford

Figure 31
Tactical: Convoy on a Narrow Road (Wunder Beach)
Photo by Keith Swilik

Figure 32
Tactical: Preparing to Fight As Infantry (Tet 68 – Company B)
Photo by Roger Collis

Figure 33
Tactical: Fight as Infantry – Digging In (Tet 68 – Company B)
Photo by Roger Collis

Figure 34
Tactical: Artillery Fire Defending LZ Jane (Common Occurrence)
Photo by Keith Swilik

Figure 35
Tactical: Mine Detection
Photo by Lawrence Doff

Figure 36
Enemy Actions: Enemy Demolition of Culvert Headwall
Photo by Jim Thacker

Figure 37
Enemy Actions: HHC Supply Room
Photo by Jim Thacke

Figure 38
Enemy Actions: Enemy Installed Air Conditioning
Photo by Jerry Nichols

Figure 39
Enemy Actions: Sniper Fire
Photo by John Miller

Figure 40
Enemy Actions: Mine Damage to Scraper
Photo by Richard Carpenter

Figure 41
Enemy Actions: Mine Damage to 5-Ton Dump Truck
Photo by Dennis Cluth

Figure 42
Enemy Actions: Mine Damage to 5-Ton Tractor
Photo by Howard Getchell

Figure 43
Enemy Actions: Mine Damage to 5-Ton Dump Truck
Photo by John Miller

Figure 44
Enemy Actions: Mine Damage to Grader
Photo by Larry Jinkins

Figure 45
Enemy Actions: Mine Damage to 2 ½ -Ton Truck
Photo by Jim Swick

Figure 46
Enemy Actions: Mine Damage to Scraper
Photo by Keith Swilik

Figure 47
Enemy Actions: Bridge Demolition
Photo by John Van Zelf

Figure 48
Bridge Construction
Photo by Dave Harbach

Figure 49
Bridge Construction
Photo by Chuck Schueddig

Figure 50
Bridge Construction: Bailey (Panel) Bridge
Photo by John Van Zelf

Figure 51
Bridge Classification by Load Testing
Photo by Keith Swilik

Figure 52
Quarry Operations
Photo by Howard Graver

Figure 53
Quarry Operations
Photo by Frank Lowe

Figure 54
Land Clearing: Rock Drilling
Photo by Dennis Cluth

Figure 55
Land Clearing: Demolition Preparation
Photo by Hank Ulman

Figure 56
Land Clearing: Rock Blasting
Photo by Frank Lowe

Figure 57
Land Clearing: Initial Dozer Entry Into An Area
Photo by Frank Lowe

Figure 58
Land Clearing: D7 Rome Plow Entering to Clear Jungle
Photo by Larry Jinkins

Figure 59
Land Clearing: Cleared Area
Photo by Dave Harbach

Figure 60
Road Construction: Culvert Placement
Photo by Chuck Schueddig

Figure 51
Road Construction: Culvert Construction
Photo by Fred Oysterman

Figure 62
Road Construction: Culvert Placement
Photo by Jim Swick

Figure 63
Road Construction: Fill Placement and Compaction (Wunder Beach Road)
Photo by Dennis Cluth

Figure 64
Road Construction: Tandem Grading for Better Security (QL-11 Song Pha)
Photo by Bill Stafford

Figure 65
Road Construction: Paving Operation
Photo by Howard Getchell

Figure 66
Road Construction: QL-19 An Khe Pass
Photo by Dave Harbach

Figure 67
Road Construction: QL-19 An Khe Pass After Construction
Photographer Unknown

Figure 68
Road Construction: QL-19
Photo by Lawrence Doff

Figure 69
Road Construction: QL-19 Mang Yang Pass Today
Photo Obtained by Larry Jinkins

Figure 70
Road Construction: Section of QL-1 in 1969
Photo by Frank Lowe

Figure 71
Road Construction: QL-1 Today
Photo Obtained by Larry Jinkins

Figure 72
Road Construction: QL-11 in 1968
Photo by Keith Swilik

Figure 73
Road Construction: QL-11 in 1968
Photo by Kieth Swilik

Figure 74
Road Construction: QL-11 in 1968
Photo by Keith Swilik

Figure 75
Road Construction: Tan My Bridge Dividing Companies B & C Operational Areas
Photo by John Van Zelf

Figure 76
Road Construction: QL-11 (NH 27) in Song Pha Today
Photo Obtained by Larry Jinkins

Figure 77
Road Construction: QL-11 (NH 27) Today
Photo Obtained by Larry Jinkins

Figure 78
Road Construction: QL -11 Good View Pass in 1968
Photo by Mark Harbert

Figure 79
Road Construction: QL-11 Good View Pass in 1968
Photo by Dennis Cluth

Figure 80
Road Construction: QL-11 (NH 27) Good View Pass Today
Photo Obtained by Larry Jinkins

Figure 81
Road Construction: QL-11 (NH 27) Good View Pass Today
Photo Obtained by Larry Jinkins

Figure 82
Port Construction
Photo by Dave Harbach

Figure 83
Airfield Construction: Building Air Cargo Storage Area @ Camp Evans
Photo by Keith Swilik

Figure 84
Air Field Construction: Placing Culvert
Photo by Dave Harbach

Figure 85
Air Field Construction: Grading and Oil
Photo by John Miller

Figure 86
Air Field Construction: Mat Placement
Photo by Chuck Schueddig

Figure 87
Air Field Construction: Song Mao Airfield (1968-71)
Photo Obtained by Larry Jinkins

Figure 88
Air Field Construction: Song Mao Airfield – Today
Photo Obtained by Larry Jinkins

Figure 89
Air Field Construction: Vinh Than Special Forces Camp
Photo by Dave Harbach

Figure 90
Building Construction
Photo by Jim Swick

Figure 91
Building Construction
Photo by John Van Zelf

Figure 92
Materials Testing: Soil Density
Photo by Chuck Schueddig

Looks like testing concrete for right formula

Figure 93
Materials Testing: Concrete Slump
Photo by Chuck Schueddig

Figure 94
Civic Action: Orphanage Support
Photo by Frank Lowe

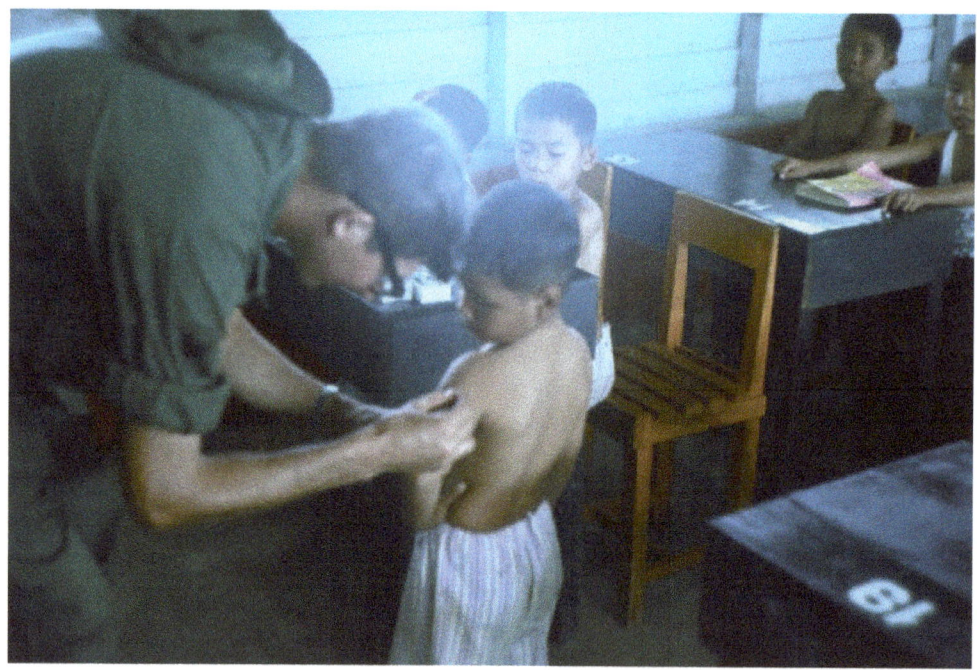

Figure 95
Civic Action: Inoculations
Photo by Dave Harbach

Figure 96
Support: Food Preparation
Photo by Jim Swick

Figure 97
Support: Communications
Photo by Richard Carlson

Figure 98
Support: Religious Services
Photo by Fred Osterman

Figure 99
Support: Equipment Maintenance
Photo by Bill Heflin

Figure 100
Support: Water Purification and Transportation
Photo by Clyde Hutson

Figure 101
"We were young once - And, Skinny"
Photo by Keith Swilik

Figure 102
"End of Mission"
Photo by Jim Thacker

www.ingramcontent.com/pod-product-compliance
Lightning Source LLC
Chambersburg PA
CBHW041632040426

42446CB00024B/3492